Reihe Göttinger Pferdewissenschaften

Band 1

Adaptation strategies of Shetland ponies *(Equus ferus caballus)* to seasonal changes in climatic conditions and food availability

Dissertation
for the Doctoral Degree
at the Faculty of Agricultural Sciences,
Georg-August-University Goettingen,
Germany

presented by
Lea Brinkmann, née Mann
born in Göttingen

Göttingen, February 2012

Bibliografische Information der Deutschen Nationalbibliothek

Die Deutsche Nationalbibliothek verzeichnet diese Publikation in der Deutschen Nationalbibliografie; detaillierte bibliografische Daten sind im Internet über http://dnb.d-nb.de abrufbar.

1. Aufl. - Göttingen : Cuvillier, 2012
    Zugl.: Göttingen, Univ., Diss., 2012

978-3-95404-209-8

D7

1. Referee:	Prof. Dr. Martina Gerken
2. Co-referee:	Prof. Dr. Dr. Matthias Gauly
Date of disputation:	February 9, 2012

TABLE OF CONTENTS

SUMMARY

The study was conducted from February 2010 until February 2011 at the Department of Animal Sciences, University of Goettingen. The objective of the study was to detect adaptative mechanisms of Shetland ponies to environmental changes in climate and feed supply. Changes in several physiological and behavioural parameters were monitored to evaluate possible welfare and health impairment in particular during the winter period.

The study started with 8 female Shetland ponies but the herd was increased to 10 mares from the beginning of May 2010. From February 2010 until May 2010 ponies were housed in two groups of four ponies each in two identical pens with permanent access to an outdoor paddock. In order to ensure that all animal would ingest the required amount of feed, both pens were equipped with 5 feeding stands each. The temperatures inside and outside the stable were comparable. All horses received the same feeding regime with straw and hay *ad libitum*, supplemented with small amounts of mineral mixture and concentrate.

From May 2010 until October 2010 all mares were kept in a group on a permanent pasture where they were fed small portions of hay, straw and a mineral supplement in addition to the grass. In October the animals were brought back into the same stable as before, where they were allocated into one control and one treatment group of five animals each, resulting in a comparable mean body condition score (BCS) and body weight (BW) for both groups. While from the beginning of November the control group was fed as before, the amount of feed offered in the treatment group was gradually reduced from 100% to 70% of the maintenance requirements for the respective Shetland pony. Water was available *ad libitum* throughout the experiment.

During the entire study BW, resting heart rate (HR) and hair length were recorded at biweekly intervals and body condition was scored monthly. The ambient temperature (T_a) was registered every 10 min.

To record the locomotor activity (LA) and lying time, Activity-Lying-Temperature-Pedometers (ALT-Pedomerter) were fitted to the ponies' legs that continuously recorded the activity impulses and the time spent lying. The subcutaneous temperature (Ts), determined every two hours, was measured by a miniature temperature data logger surgically implanted in the animals' neck prior to the study.

Total body water and total water intake were measured using the D_2O dilution technique by monthly application of the stable isotope Deuterium. From August 2010 until the end of the study blood samples were obtained from the mares each month. Non-esterified fatty acids (NEFA), β-hydroxybutyrat (BHB), total protein (TP), thyroxin (T_4) and total bilirubin (TB) were analysed. Additionally, the rectal temperature was recorded biweekly during the feeding trial from November 2010 until February 2011.

Body weight, BCS, HR and LA were highly correlated with T_a, showing high values in summer (average means ± SD of 160.8 ± 38.4 kg, 3.9 ± 1.0 BCS, 63.2 ± 7.6 beats/min and 23,100 ± 4140 activity impulses, respectively) and low values in winter (average means ± SD of 137.9 ± 35.9 kg, 3.1 ± 0.7 BCS, 32.6 ± 4.6 beats/min and 7,390 ± 1070 activity impulses, respectively). The LA showed biphasic diurnal rhythms with highest values around dawn and dusk. Mean daily lying times were highest in spring (March: 169 min) and autumn (August: 127 min) and low in winter (December: 56 min). Resting appeared usually at night, with the highest lying times observed before dawn.

The mean T_s measured over the entire year was 35.6 ± 05 °C. Daily mean T_s showed considerable fluctuations during spring and summer in all animals. However, these high variations decreased during autumn and winter. Daily minimum T_s was with 28.2 °C lowest in April and with 35.7 °C highest in October. Lowest daily T_s typically appeared in the early morning hours around dawn and increased with early daylight.

The total body water (%) of the ponies averaged 64.9 ± 4.0% with no differences between seasons, whereas the total water intake was highest in summer (135 ml/kg BW) and decreased by 60% in winter (51 ml/kg BW). The multiple regressions calculated between total water intake and T_a, HR, LA and BW explained 90% of the variation in total water intake, however, only T_a and HR had a significant influence on water intake. Hair length was significantly shorter in summer compared to winter for all animals.

During the grazing period in summer the animals did not differ significantly in their blood parameters. However, concentrations of TP, BHB and T_4 were significantly higher in summer than in winter, while the reverse was observed for NEFA and TB concentrations.

While the restrictive feeding of one group for four months had no significant effect on total body water, total water intake and LA, the BCS, BW, rectal temperature and HR decreased continuously in the restricted group. Furthermore, the T_s was significantly lower in the treatment group on 52 of the 120 feed restricted days compared to the control animals. The guard hair length was significantly longer in the restricted animals compared to the control animals at the end of the feeding trial. Feed restricted animals showed lower TP and BHB, but higher NEFA and TB concentrations compared to the control group. T_4 concentrations tended to be lower in the feed restricted mares.

The ponies adapted to the low temperatures in winter presumably by a lower energy expenditure due to reduced LA. Furthermore, they reduced their lying times on the cold soil and increased their hair length in winter to limit the heat loss. The lowered HR and the reduced T_4 concentrations in winter indicate a reduction in the metabolic rate to save body energy reserves.

The long-term recording of LA and T_s revealed that the ponies not only saved energy in winter, but exhibited a daily hypometabolism over the entire year. Thus, the animals reduced their activity and allowed the body temperature to fall during the night to a nadir in the morning. These adaptive mechanisms resulted in a reduced heat loss in winter nights due to a lower temperature gradient between body and environment, while the same mechanism allowed saving energy from heat dissipation in summer due to higher absorption capacity of heat during the day.

During the feed shortage under winter conditions the restricted animals reduced their metabolic rate even further as indicated by continuously decreasing HR, T_4 concentrations, rectal temperature and the lower T_s compared to the control animals. However, the metabolic reduction was not sufficient to compensate for the lower food energy availability in restricted animals. Thus, animals mobilized body fat to cover their energy demand, resulting in changes in blood parameters. In particular levels of total protein and total bilirubin were outside the critical values, indicating beginning health problems especially with regard to the liver metabolism.

The comparison of the feed restricted Shetland ponies with other feed restricted or starved herbivores showed that these species showed similar reactions. Contrary to speculations, that domestic animals lost their adaptive mechanisms for adaptation to harsh environments compared to wild species, the Shetland ponies showed very effective mechanisms for fat mobilization and metabolic reduction during feed shortage. Similarly, the observed daily hypometabolism was comparable to that detected in some wild Northern herbivores.

The Shetland ponies adapted effectively to the changing environmental conditions by several behavioural and physiological mechanisms so that the continuous outdoor housing can be an adequate housing system. However, impairment of welfare and health is expected, when horses are kept in winter outdoor housing without additional feed supply.

ZUSAMMENFASSUNG

Ziel der Studie war es, Anpassungsmechanismen von Shetlandponys an klimatische Veränderungen und Änderungen in der Futterverfügbarkeit zu untersuchen. Des Weiteren sollte die Adaptation an Futterknappheit, wie sie bei sehr extensiver Haltung unter natürlichen Verhältnissen im Winter zu finden ist, untersucht werden. Der Versuch wurde von Februar 2010 bis Februar 2011 am Department für Nutztierwissenschaften der Georg-August-Universität Göttingen durchgeführt und begann mit acht weiblichen Ponys, wurde aber im Mai 2010 um zwei Stuten ergänzt. Die Tiere wurden von Februar 2010 bis Mai 2010 in zwei Gruppen à vier Tieren auf Paddocks mit Zugang zu identisch gestalteten Stallabteilen zu gehalten. Beide Abteile waren mit Fressständen ausgestattet, um zu gewährleisten, dass jedes Tier die ihm zugeteilte Menge Futter aufnehmen konnte. Die Stalltemperaturen waren vergleichbar mit den Außertemperaturen. Ab Mai 2010 wurden alle Ponys zusammen auf einer Dauerweide gehalten und im Oktober 2010 erneut in den oben beschriebenen Stall eingestallt. Dabei wurden die Ponys in eine Kontrollgruppe und eine Versuchsgruppe à 5 Tiere aufgeteilt, so dass beide Gruppen vergleichbare Körperkonditionen (Body Condition Scores, BCS) und Körpergewichte (KG) aufwiesen.

Alle Tiere wurden von Februar 2010 bis Mai 2010 mit Stroh und Heu *ad libitum* gefüttert. Zusätzlich wurde den Tieren Mineralfutter und eine geringe Menge an Kraftfutter zur Verfügung gestellt. Von Mai 2010 bis Oktober 2010 standen den Tieren zusätzlich zum Weideaufwuchs geringe Mengen an Stroh, Heu und Mineralfutter zur Verfügung. Während des im November beginnenden Fütterungsversuches erhielt die Kontrollgruppe das gleiche Futter wie zu Beginn der Studie, die Futtermenge der Versuchsgruppe wurde schrittweise von 100 auf 70% des Erhaltungsbedarfs des jeweiligen Shetlandponys herabgesetzt.

Während des gesamten Versuchs wurde das KG, der Ruhepuls (HR) und die Haarlänge in zweiwöchigen Intervallen gemessen, der BCS wurde monatlich erfasst und die Umgebungstemperatur täglich alle 10 Minuten. Um die Bewegungsaktivität (LA) zu erfassen, wurden ALT-Pedometer an den Beinen der Tiere befestigt, die kontinuierlich die Aktivitätsimpulse und die Liegezeit aufzeichneten. Den Ponys wurden Miniatur Datenlogger im Halsbereich unter die Haut implantiert, die zweistündlich die subkutane Temperatur erfassten. Des Weiteren wurden der

Körperwassergehalt und die Gesamtwasseraufnahme der Tiere durch die Isotopen-Verdünnungsmethode ermittelt. Dafür wurde den Tieren in einem vierwöchigen Rhythmus Deuterium gespritzt und die Konzentrationsabnahme im Körper ermittelt. Zusätzlich wurde den Ponys von August 2010 bis Februar 2011 monatlich eine Blutprobe entnommen und die Konzentrationen an Nicht-Veresterten-Fettsäuren (NEFA), Total-Bilirubin (TB), Total-Protein (TP), β-Hydroxybutyrat (BHB) und Thyroxin (T_4) untersucht.

KG, BCS, LA und HR korrelierten stark mit der Umgebungstemperatur und zeigten hohe Werte im Sommer (Mittelwerte ± SD: 160,8 ± 38,4 kg; 3,9 ± 1,0 BCS; 63,2 ± 7,6 Schläge/Min und 23.100 ± 4140 Aktivitätsimpulse) und niedrige im Winter (Mittelwerte ± SD: 137,9 ± 35,9 kg; 3,1 ± 0,7 BCS; 32,6 ± 4,6 Schläge/Min und 7.390 ± 1070 Aktivitätsimpulse). Im Tagesrhythmus war die LA stark von der Photoperiode abhängig und zeigte höchste LA kurz nach Sonnenauf- und -untergang. Erhöhte Liegezeiten waren im Frühjahr und im Herbst, geringste im Winter zu beobachten. Die subkutane Tagesdurchschnittstemperatur wies starke Schwankungen im Frühjahr und in Sommer auf, welche aber zum Herbst und Winter abnahmen. Die mittlere tägliche Minimaltemperatur war mit 28,8 °C im April am geringsten und mit 35,7 °C im Oktober am höchsten. Tägliche Tiefstwerte der subkutanen Temperatur traten typischerweise am frühen Morgen vor Sonnenaufgang auf, gefolgt von einem starken Anstieg der Temperatur über Tag.

Der Körperwassergehalt der Tiere hatte einen Mittelwert von 64,9 ± 4,0% und zeigte keinen Unterschied zwischen den Jahreszeiten. Im Gegensatz dazu zeigten sich Höchstwerte in der Wasseraufnahme von 135 mL/kg KG im Sommer und um 60% reduzierte Werte von 51 mL/kg KG im Winter. Allerdings konsumierten alle Ponys im Dezember ungewöhnlich hohe Mengen an Wasser. Die multiple Regression zwischen Wasseraufnahme, LA, HR und Umgebungstemperatur erklärte 90% der Variation in der Wasseraufnahme, wobei die Wasseraufnahme nur durch die Umgebungstemperatur und die HR signifikant beeinflusst war. Die gemessene Haarlänge des Fells war im Sommer deutlich kürzer als im Winter. Die Konzentrationen der Blutparameter TP, BHB und T_4 im Sommer waren höher, die an TB und NEFA geringer als im Winter.

Während die Futterrestriktion keinen Einfluss auf den Körperwassergehalt, die Wasseraufnahme und die LA zeigte, kam es zu einer kontinuierlichen Abnahme der HR, der Rektaltemperatur, des KG und des BCS. Zusätzlich zeigte die

Versuchsgruppe eine geringere subkutane Temperatur an 52 von 120 restriktiv gefütterten Versuchstagen. Das Oberhaar der restriktiv gefütterten Tiere war am Ende des Versuchs signifikant länger als das der Kontrollgruppe. Die Futterrestriktion führte ebenfalls zu einer Reduktion der TP und BHB, und zu einer Erhöhung der NEFA und TB Konzentrationen. T_4 Konzentrationen zeigten einen abnehmenden Trend.

Durch eine Verminderung der LA im Winter benötigten die Tiere in der kalten Jahreszeit weniger Energie. Zusätzlich limitierten die reduzierte Liegezeit auf kaltem Boden und die Ausbildung eines langen Haarkleides die Wärmeverluste. Verminderte HR und T_4 Konzentrationen im Winter deuten auf eine abgesenkte Stoffwechselaktivität hin, um Körperenergiereserven zu sparen. Im Sommer deuteten die erhöhte HR und Wasseraufnahme auf eine Anpassung an Hitzestress hin. Allerdings zeigten die tagesrhythmischen Veränderungen in der LA und der Ts, dass die Tiere nicht ausschließlich im Winter Energie sparten, sondern das ganze Jahr über einen täglichen Hypometabolismus zeigten. Dabei reduzierten die Ponys ihre LA und erlaubten eine Absenkung der Körpertemperatur über Nacht bis zu einem Tiefpunkt am Morgen. In kalten Nächten führte dies, durch einen reduzierten Temperaturgradienten zwischen Körper und Umgebung, zu verringerten Wärmeverlusten. Gleichzeitig führte die nächtliche Abkühlung an Sommertagen, durch die erhöhte Wärmeaufnahmekapazität des Körpers, zu einem geringeren Energieaufwand für Wärmeabgabe.

Während der Futterrestriktion im Winter reduzierte die Versuchsgruppe ihren Stoffwechsel zunehmend, was sich durch die kontinuierliche Reduktion der HR, Rektaltemperatur, subkutane Temperatur und T_4 Konzentrationen im Vergleich zur Kontrollgruppe zeigte. Allerdings erwies sich die Stoffwechselreduktion als unzureichend, um die fehlende Energieversorgung zu kompensieren. Die Tiere mobilisierten daher Körperfett, woraus Veränderungen der Blutparameter resultierten, die auf beginnende gesundheitliche Probleme, besonders die Leber betreffend, hinwiesen. Die restriktiv gefütterten Ponys zeigten im Vergleich mit anderen restriktiv gefütterten Tierarten zeigte weitgehend ähnliche Reaktionen, allerdings mit einigen spezifischen Besonderheiten.

Der beobachtete tägliche Hypometabolismus der Ponys war vergleichbar mit dem, der bei einigen wilden nördlichen Herbivoren gefunden wurde. Unsere Shetlandponys passten sich durch Veränderungen im Verhalten und in physiologischen Parametern erfolgreich an sich verändernde klimatische Bedingungen an, so dass eine ganzjährige Außenhaltung der Tiere als tiergerecht angesehen werden kann. Auch wenn die Ponys Anpassungsmechanismen an die Futterknappheit in Winter zeigten, sind gesundheitliche Folgen für die Tiere bei einer Winteraußenhaltung ohne Zufütterung zu erwarten.

CHAPTER 1

INTRODUCTION

INTRODUCTION

An increasing number of domesticated horses are kept in outdoor group housing systems. These systems are ascribed to allow the performance of the horses' natural behaviour patterns. However, in Central Europe ambient temperatures can easily range from -25 °C to +35 °C over the course of the year, thus exposing animals to extreme environmental conditions. Horses need to adapt to the varying external effects by changes in physiological parameters, behaviour and metabolism to sustain their homeostasis (Arnold and Dudzinski, 1978). Wild horses are known to adapt by accumulation of body fat, reduced activity and a reduced metabolic rate (Scheibe and Streich, 2003; Arnold et al., 2006; Berger et al., 2006), while domesticated horses may have a reduced adaptive capacity so that especially in winter the environmental conditions can result in health problems. Little information is available on the adaptation of domestic horses under such conditions. Furthermore, there is limited knowledge if the extensive outdoor housing can be described as an appropriate housing system for horses.

Adaptation is a widely used term that describes the modification processes during the evolution, the ontogenesis and the actual genesis as well as the resulting modification of body structure, physiology and behaviour. Behavioural adaptation is based on genetics, learning, formation of memory, memory and tradition creation (Gattermann, 2006). The adaptation ability of domesticated horses may be reduced as the transfer of the animal to the human household represents a manipulation of the natural selection. The human being changed the animals' environment in such a manner that the struggle for existence, decisive for natural selection, lost importance. Furthermore, the animal was modified by zootechnical measures. The artificial selection by humans resulted in changes of body structure, organic functions and behaviour (Scheunert and Trautmann, 1987). Nevertheless, behaviour patterns that were carried over from the ancestors but lost their function or their primary role still exist as behavioural relics (Bogner and Grauvogl, 1984). In general, domesticated horses are gentler, less aggressive and have reduced sensual performance (Zeitler-Feicht, 2001).

Domestic horses can be subdivided in three subcategories: ponies, cold-blooded horses and warm-blooded horses. The cold-blooded horses are massive, heavy, short built and have a high compactness index (Langlois, 1994; Zeitler-Feicht, 2001).

Warm-blooded horses are high, long-limbed and slender with long extremities (Langlois, 1994) and show less body weight than the cold-blooded horses (Zeitler-Feicht, 2001). In contrary ponies have the tendency for restricted growth, round shape, short extremities and opulent mane and tail hair. Their hair coat is longer and they have more undercoat especially in winter. For this reason they are suited for loose housing. Ponies also have a high feed conversion and tend towards adiposity (Zeitler-Feicht, 2001).

Thermoregulation as adaptive mechanism in endothermic animals

Under extensive housing conditions, among the different adaptive mechanisms, thermoregulation is of special importance for horses. Their body temperature needs to be kept in narrow limits to ensure the reactivity and to sustain the function of the brain. This chapter will focus on their thermoregulatory mechanisms.

The horse belongs to the homoeothermic animals. These animals possess a highly developed temperature regulation and are able to keep their metabolic rate and body temperature relatively constant with changing environmental temperatures (Crompton et al., 1978; Bianca, 1979; Ousey et al., 1992; Singer, 2007). This enables these animals to sustain a constant metabolic rate and reactivity in the vital organs over a wide climatic range. The price for that are a four- to eightfold increased specific energy demand and the need for continuous substrate supply (Else and Hulbert, 1981; Singer, 2007).

The body of homoeothermic animals can be subdivided in the body shell and the body core. The temperature of the body core (organs, thoracic- and abdominal cavity, brain) is kept relatively constant by thermoregulatory mechanisms, the body shell temperature lies below the core temperature and varies depending on thermoregulatory needs. The body shell is composed by the skin, hypoderm and temporary by the muscles and extremities. There are high topographical temperature differences in the body shell. Horses distal limb temperatures can be reduced to 1.7 °C (Palmer, 1983), whereas the skin of the torso shows high temperatures with only slight fluctuations (Scheunert and Trautmann, 1987; Penzlin, 1991; Schmidt-Nielsen, 1997).

As changes in heat loss due to changes in ambient temperature, humidity and wind speed and heat production e.g. by activity and feed consumption may occur quickly, the body needs to possess efficient mechanisms for the control of heat loss and heat production (Scheunert and Trautmann, 1987). Reactions regarding thermal variations can be functional, structural or ethological (Bianca, 1977). Functional and structural changes can be characterised as physiological temperature regulation (Wollenweber, 2007). A survey on physiological, chemical and behavioural mechanisms of thermoregulation is given in Tab. 1.

Tab. 1 Physical, chemical and behavioural measures of thermoregulation (Wittke, 1972).

Climate	Measures for temperature regulation		
	Autonomic regulation		Behavioural regulation
	Physical	Chemical	
Heat	Increased blood circulation of the skin	Decrease in heat production	e.g. Searching for shade, wetting with water
	Panting, Sweating		Reduction in activity and food consumption
Cold	Reduced blood circulation of the skin	Increase in heat production	e.g. Grouping (collective temperature regulation),
	Piloerection		searching for wind shadow

Surface receptors in the skin as well as thermoreceptors in the body core submit continuously terminal impulses to the hypothalamus (Hensel, 1966) where the received information is compared with an internal `set-point`. The hypothalamus determines if the body temperature is too cold, too hot or if the temperature is appropriate (Sjaastad et al., 2003). If the body temperature deviates from the `set-point`, different thermoregulatory mechanisms are implemented (Hensel, 1966).

Body heat, generated by metabolic waste heat, must equal the heat loss to sustain homoeothermic (Scheunert and Trautmann, 1987; Schmidt-Nielsen, 1997).

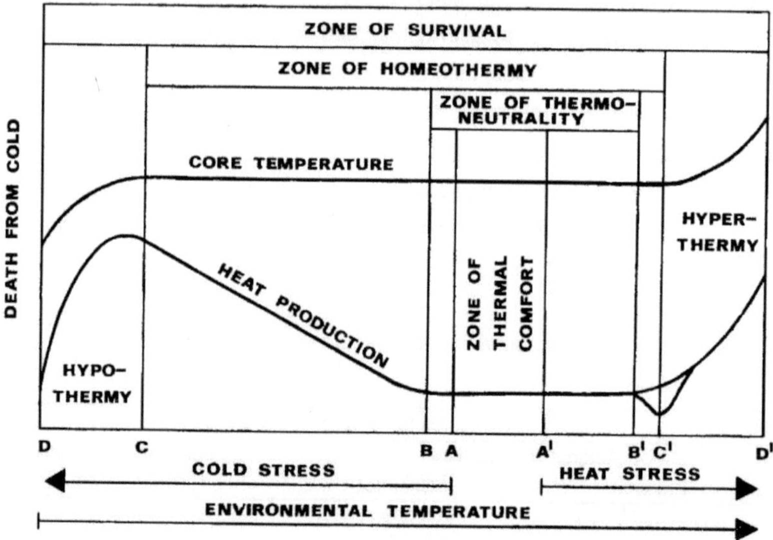

Fig. 1 Schematically illustration of the thermoneutral zone and the critical temperatures after Bianca (1968).

Bianca (1968) schematically outlined the reaction of the homoeothermic animal to changes in the ambient temperature (Fig.1). The thermoneutral zone (B - B`) is the temperature zone in which the generated heat from the basal metabolic rate is sufficient to keep the body temperature constant without larger thermoregulatory mechanisms (Bianca, 1979; Scheunert and Trautmann, 1987). For horses there are highly variable data for this temperature zone. McBride (1985) states -15 to 10 °C, Clarke (1987) 10 to 30 °C while Morgan (1997a) found a temperature zone of 5 to 25 °C. However, these specifications present guideline values that can be influenced by factors like humidity, wind speed, precipitation and radiation (Schrader, 2009). If the ambient temperature decreases below a certain limit, the lower critical temperature (B), more heat is required to compensate the heat losses and therefore the metabolic rate is increased (Schmidt-Nielsen, 1997). Above the upper critical temperature (B`) the metabolic rate and likewise the heat loss increases due to sweating and panting. If the heat loss mechanisms become insufficient in hot and humid conditions, the body temperature increases until death of heat occurs at body temperatures of 43-44

°C, primarily because of denaturation of certain enzymes and regulatory proteins (Sjaastad et al., 2003). On the other hand death of cold may occur when the heat production in the body cannot compensate the heat losses due to extreme low temperatures (Bianca, 1971) or insufficient feed supply.

Therefore, the seasonal feed shortage in the temperate zone in the winter months is standing in contrast to the higher thermoregulatory energy demand in winter. The homoeothermic animals adapted to this dilemma by migration, the creation of feed cache, higher body insulation or the reduction of heat loss by decreasing the body temperature and hibernation (Bligh, 1998; Singer, 2007).

Ethological/behavioural thermoregulation

Changes in the animals' behaviour are important thermoregulatory mechanisms (Hetem et al., 2007). The main function of the behavioural thermoregulation is the minimization or prevention of heat loss (Bianca, 1977). Behavioural thermoregulation can be divided in heat producing and heat retraining mechanisms (Autio, 2008) and means the selection of an environment and activity that reduces the thermal stress. Behavioural thermoregulation requires less energy than thermoregulation by sweating or by burning calories to maintain the core body temperature (Houpt, 2005). Therefore, it is initiated before physiological thermoregulation occurs (Bianca, 1977).

In a cold environment the main mechanisms of the behavioural thermoregulation are the reduction of the relative body surface (Bianca, 1977), shelter- and comfort seeking (Hetem et al., 2007) and turning the smallest surface towards the wind (Boyd and Houpt, 1994). A reduction of the morphological body surface is not possible but a reduction of the freely exposed, heat dissipating surface by grouping and standing close to each other is possible (Bianca, 1979; Langlois, 1994). The time spent lying is reduced (Duncan, 1985) to avoid heat loss to the ground via conduction. The animal seeks shelter and comfort in warmer, sunny microclimates and weather shielded areas (Bianca, 1979) such as trees, bushes, and lower elevations (Zeeb, 1994). Furthermore, horses reduce their locomotor activity to preserve energy reserves and increase the time spent grazing to maximise their energy intake as caloric needs are greater in the cold (Duncan, 1980; 1985; Berger et al., 1999; 2006).

During hot weather horses search for cool, humid soil (Bianca, 1979) and breezy locations (Zeeb, 1994). They avoid direct solar radiation by standing in the shade (Crowell-Davis, 1994) and moisten the body surface with water to cool the body by evaporation (Bianca, 1977). Additionally, the horses decrease their activity and feed intake to reduce the metabolic heat production (Bianca, 1977; 1979).

Physiological and chemical thermoregulation

The physiological thermoregulation comprises structural as well as functional mechanisms. The structural modifications involve vasoconstriction and vasodilatation, increase of the coat density and coat length as well as piloerection (erection of the guard hair) (Kolb, 1967). If the body core temperature of the animal is elevated, the tone of the muscle cells in the arterioles decreases and as consequence the cutaneous blood flow and skin temperature increases. This vasodilation therefore raises the heat loss from the body since this is dependent on the temperature gradient between the skin and the environment. Vasoconstriction occurs during exposure of the animal to a cold environment. The sympathetic nervous system controls the constriction of the vessels that supply the cutaneous tissues, especially the ears and the extremities with blood. The blood flow in the dermis diminishes and the heat transfer is reduced. At a temperature around 0 °C vasodilation occurs by sudden opening of the vessels to permit intermitted warming of the skin (Radostits et al., 2005). Vasoconstriction was observed by several authors in horses held under cold conditions (Palmer, 1983; Mogg and Pollitt, 1992; Morgan, 1997b).

The horses' hair coat density and length is dependent on the ambient temperature, photoperiod and horse breed (Kooistra and Ginther, 1975; Cymbaluk, 1990; Zeitler-Feicht, 2001) and it serves to create a static air layer above the skin (Penzlin, 1991) to reduce the heat loss. Cold-blooded horses and ponies have a higher coat density and length than Arabians and Thoroughbreds (Mills and McDonnell, 2005). Zeitler-Feicht (2001) describes a more dense short hair and a rough guard hair in ponies. However, cold housed horses do not inevitably grow a thick winter coat when they are adequately fed (McBride et al., 1985).

Each guard hair follicle has a muscle, called piloerector muscle *(arrector pili)*, which is located in the outer layer of the dermis and attached to the hair bulb. An increase in the impulse frequency of the sympathetic nerves causes contraction of the muscle. Thereby the hair is erected and the distance between the skin and the tip of the hairs increases. This increases the thickness of the air layer between the hairs used for body insulation (Sjaastad et al., 2003). Ousey (1992) showed an increase of the coat depth by 0.3 to 1.4 cm in foals and Young and Coote (1973) reported of a 20-30% higher hair coat depth in adult horses. Piloerection on different parts of the body can have various effects. A complete piloerection on the dorsal surfaces actually might increase the heat loss since warm air rises, while piloerection on the ventral surfaces conserves heat (Davenport, 1992).

Nearly all mentioned structural mechanisms serve to reduce the heat loss. The heat dissipation under a hot environment is mainly induced by functional thermoregulation. The animals' body heat can be dissipated by radiation, conduction, convection and transpiration (Fig. 2).

Radiation

Warm bodies exchange heat with cooler bodies by radiation (Langlois, 1994). The heat dissipated by radiation is dependent on the temperature difference between the body surface and the environment (Scheunert and Trautmann, 1987). In general, the animal's body has a higher temperature than the environment and therefore looses heat (Langlois, 1994). Horses can reduce the heat loss trough radiation by reducing the skin temperature. On the contrary, the increase in skin temperature by increased peripheral blood circulation results in a higher heat loss by radiation (Scheunert and Trautmann, 1987). The cooling by radiation may be high at night during cold winter month (Cymbaluk and Christison, 1989), but horses can reduce the heat loss by standing close to each other and by seeking shelter (Bligh, 1998; Mejdell and Boe, 2005).

Conduction

Conductive heat loss takes place by the direct contact with objects and is likewise dependent on the temperature gradient between the animals and the environment (Langlois, 1994). In a standing position the horse is only loosing small amounts of heat by conduction, while during lying this heat loss can be remarkable (Clark, 1994). In summer horses will therefore seek for cooler soil to lie down. Wet body- or/and lying surfaces will enhance the heat loss by conduction as water has a high thermal conductivity (Autio, 2008). Thus, in winter horses reduce the lying times on the cold and humid soil to minimize heat loss (Duncan, 1985; Langlois, 1994).

Convection

Convection is the heat transfer via moving air or water and depends on the temperature gradient. If the body temperature is higher than the ambient temperature, the air that is in contact with the skin will be heated by conduction. The heated air ascents and is replaced by cooler air (Sjaastad et al., 2003). This free convection elevates with decreasing temperatures (McArthur, 1991) but is of limited significance for the horse as it is equipped with a fur. Forced convection due to wind is of greater importance (Sjaastad et al., 2003). To minimize heat loss in winter horses are seeking for wind protected areas. In summer during high temperatures they are looking for windy places to enhance this type of heat loss (Zeeb, 1994). Convection also occurs within the body. This means that the heat generated in the muscles and organs is transferred to the skin by the blood (Vogel, 2006).

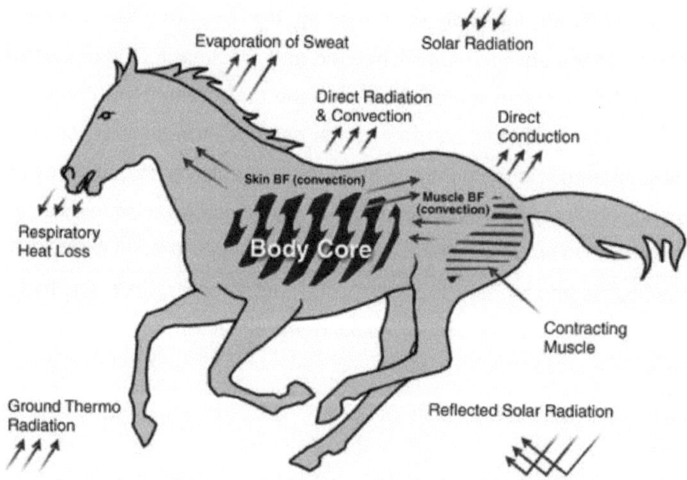

Fig. 2 Functional mechanisms of heat dissipation to the environment and heat gain by the environment in the horse (BF: blood flow) (Wood, 2009).

Evaporation

The evaporation is the most efficient way to loose body heat under hot conditions (Scheunert and Trautmann, 1987) and is associated with the loss of water vapour from the respiration system and the skin (Clark, 1994). There are 2430 kJ detracted by the evaporation of one litre water on the body surface (Scheunert and Trautmann, 1987). With high ambient temperatures the heat loss via evaporation in horses increases while the loss by non-evaporative mechanisms decreases. However, evaporation depends on the difference in water vapour pressure between the skin and the ambient air and the amount of exhaled air (Morgan et al., 1997).

Small amounts of water that diffuses continuously trough the skin and evaporates are described as insensible evaporation. This water loss cannot be controlled as water molecules diffuse continuously through the skin. At low ambient temperatures the insensible evaporation stays almost constant and accounts for 20% of the entire heat loss (Morgan, 1996). The sensible evaporation includes panting, sweating and salivation whereupon sweating is of higher importance for horses (Scheunert and Trautmann, 1987; von Engelhardt and Breves, 2000) even if water is also lost by respiration.

Chemical thermoregulation

In the homoeothermic animal there are three chemical mechanisms to generate heat which all depend on the metabolic reactions involved in synthesis and breakdown of ATP (Langlois, 1994). These three ways are the basal metabolic rate, the shivering thermogenesis and the non-shivering thermogenesis (Cannon and Nedergaard, 2004). The basal metabolic rate is the rate of energy metabolism measured of a resting, awake, fasting animal in the thermoneutral zone. Thus, the basal metabolic rate is needed to maintain cell and organ functions (Bligh, 1998). Sixty percent of the body's heat production during rest is produced by heart, liver, kidneys and brain (Sjaastad et al., 2003); the remaining heat is generated by muscles, skin and skeleton (Autio, 2008). The metabolic rate rises with hormonal factors and muscular contraction (Irvine, 1967; Langlois, 1994).

The shivering thermogenesis refers to involuntary tonic and rhythmic contraction of both flexor and extensor at the same time and occurs as a reflex (Scheunert and Trautmann, 1987; Langlois, 1994). Though shivering is usually an acute response to abrupt cold exposure (Cannon and Nedergaard, 2004; Radostits et al., 2005) and increases the heat production of the animal to about four times of the basal metabolic rate (McArthur, 1991). However, the shivering thermogenesis increases the blood circulation in the extremities and results in an elevated heat loss and energy demand (von Engelhardt and Breves, 2000). Shivering was shown by Irvine (1967), McBride (1985), Morgan (1997a) and Morgan et al. (1997) in mature horses, by Mejdell and Boe (2005) for Icelandic horses and by Ousey et al. (1992) for pony foals exposed to cold temperatures.

The non-shivering thermogenesis via brown adipose tissue is more economical (Kolb, 1967). It is a very particular process that enables the brown adipose tissue to produce heat. Energy, released by the oxidation of glucose and fat, is directly dissipated as heat instead of being synthesized to ATP (Langlois, 1994). Brown adipose tissue exists in small body sized mammalian species (Dawkins and Hull, 1964; Foster and Frydman, 1978), but also in newborns of larger species (calves, lambs) (Alexander et al., 1975; Symonds et al., 1992). However, in the foal, there is no evidence of any brown adipose tissue stores (Acworth, 2003). Adult larger mammals lack brown adipose tissues (von Engelhardt and Breves, 2000). Thus, this thermogenesis is irrelevant for juvenile and adult horses.

Torpor and hibernation: a specific thermoregulatory and metabolic adaptation

In cold winters the energy demand of the animals to maintain a high body temperature is high while the vegetation which provides the primary energy supply is scarce in temperate regions (Bligh, 1998). Especially small animals have high metabolic rates that would increase further at low ambient temperature and become too expensive when feed is rare (Schmidt-Nielsen, 1997). A solution to solve this problem is to let the body temperature fall towards the environmental temperature (Bligh, 1998) and reduce the energy expenditure (French, 1982) by entering daily torpor or hibernation. It is assumed is that hibernation is more related to the seasonal feed shortage than to direct cold stress (Bligh, 1998). Furthermore, hibernation and torpor are well-regulated physiological states and not comparable with temperature reduction in cold-blooded animals (Schmidt-Nielsen, 1997).

Torpor and hibernation are characterized by the controlled reduction of body temperature and metabolic rate (Bligh, 1998; Geiser, 2004) together with other physiological functions like heart rate and respiration rate (Schmidt-Nielsen, 1997). Nevertheless, it has to be differentiated between prolonged torpor or hibernation in hibernators and daily torpor in daily heterotherms (Geiser, 2004). Hibernation usually begins in autumn and lasts until spring (Heldmaier et al., 2004), however, it consists of a sequence of torpor bouts. These bouts, lasting for several days or weeks, are interrupted by periodic rewarming and normothermic periods with durations of mainly less than a day (Twente et al., 1977; French, 1982; Heldmaier et al., 2004). Hibernating animals, usually weighing between 10 – 1000 g, can reduce their body temperature by more than 35 °C (Kayser, 1961). The entire mass range of hibernators is about 5 g to 80 kg. However, reductions of the body temperature below 10 °C are restricted to species with a body weight of less than 10 kg (Geiser, 2004). Many hibernators increase their body fat content before entering the hibernation and refuse to hibernate when fat reserves are not adequate (Geiser, 2004).

Daily torpor of daily heterotherms always lasts less than a day and is interrupted by daily foraging and feeding (Geiser, 2004). Both the metabolic rate and the body temperature are maintained at a higher rate (MacMillen, 1965; Morhardt, 1970; Geiser and Ruf, 1995). The daily torpor is less seasonal than hibernation but mainly occurs in winter times (Kortner and Geiser, 2000a). Furthermore, no extensive fattening is exhibited by daily heterotherms so that they enter torpor with low body weights (Kortner and Geiser, 2000b) and the main energy supply remains consumed feed rather than body fat (Geiser, 2004).

Apart from hibernation and daily torpor many other tachymetabolic heterotherms undergo periods of hypometabolism, whereas an alleviated form can be observed regularly during the circadian rhythm (Heldmaier et al., 2004). Comparable rhythms in body temperature variation were also observed in horses with decreasing body temperatures during the night and highest temperatures in the late evening. The range of deviation in horses was 1 °C (Piccione et al., 2002) being in the range of 0.5 – 2 °C stated for other species (Heldmaier et al., 2004).

Hypometabolism as a strategy to cope with feed shortage under winter conditions was expected to be absent in large mammals. Instead fattening, increased insulation and reduced locomotor activity were assumed to be sufficient to withstand the winter (Arnold et al., 2006). Despite this, large mammals may also show short phases of deeper hypometabolism with only small changes in body temperature because of their low thermal conductance and high thermal inertia due to the large body size

(Heldmaier et al., 2004). A nocturnal hypometabolism was already detected in red deer (*Cervus elaphus*) (Arnold et al., 2004), moose (*Alces alces*) (Renecker and Hudson, 1985), ibex (*Capra ibex ibex*) (Signer et al., 2011), and Przewalski horses (*Equus ferus przewalski*) (Arnold et al., 2006)

Specific physiological changes in the horse

Generally, there is little information available concerning the adaptation of horses to cold ambient temperature and especially concerning the adaptation to long term heat stress without physical activity of the animal.

In the following, a review is given on the specific physiological changes described for horses exposed to high or low ambient temperatures.

Rectaltemperature / Skintemperature

The measurement of the rectal temperature is suited to evaluate the body core temperature that is specified in adult horses as 37.5 – 38.0 °C (Langlois, 1994; Knickel et al., 2002; Hines, 2004). Foals have a slightly higher body temperature of 37.5 – 38.5 °C (Kolb, 1967; Knickel et al., 2002). It is considered that deviations of more than 1 °C of normothermia leads to discomfort, and that a decrease of more than 10 °C or an increase of more than 5 °C is fatal (Langlois, 1994).

Exposing the horse to a hot environment results in an increasing rectal temperature (Honstein and Monty, 1977; Ott, 2005). Likewise the skin temperature raises (Honstein and Monty, 1977) as the blood circulation of the body periphery increases by vasodilation. The temperature difference between skin and environment becomes greater and the heat dissipation by radiation and conduction increases (Scheunert and Trautmann, 1987). Short-term exposure (1 hour) of horses to ambient temperatures of -3 to 37 °C did not lead to changes in the rectal temperature (Morgan, 1997).

Decreasing ambient temperatures did not result in a uniform development of the rectal temperature. In neonates the rectal temperature remained constant during exposure to cold for several hours (Ousey et al., 1992), however, cold-habituated yearlings showed a lowered rectal temperature of 0.4 °C (Cymbaluk and Christison, 1993).

Heart rate

There is variable Information concerning changes of the heart rate under different climatic conditions. According to Cymbaluk and Christison (1993) yearlings did not change their heart rate with lower ambient temperatures. Likewise Morgan (1997a) found no effect of a cold environment on the heart rate in adult horses. Similar results were shown by Art and Lekeux (1988) for ponies. On the contrary, Ousey (1992) reported increasing heart rates in foals with ambient temperature falling below 20 °C. According to Arnold et al. (2006) a long-term exposure of Przewalski horses to a cold environment led to decreased heart rates.

An increase in cardiac output was shown by McConaghy (1996) for ponies at ambient temperatures of 41 °C. At ambient temperatures above the thermoneutral zone the increasing blood circulation of the skin (Scheunert and Trautmann, 1987) requires a higher cardiac output per minute so that the cardiac output as well as the heart rate increases with increasing ambient temperature (Kolb, 1967). However, Morgan (1997a) as well as Art and Lekeux (1988) found no change of the heart rate with higher ambient temperatures.

Respiration rate

With higher ambient temperatures the heat loss via evaporation needs to increase, since the temperature gradient between the skin and the environment is lowered and the non-evaporative heat loss is reduced. Not only the heat loss by sweating increases but likewise the heat loss from the respiratory system rises (Morgan, 1997a). With a higher in- and exhalation the amount of water lost and with it the amount of heat that can be dissipated increases. Few data is available for changes in respiration rate of horses under changing environmental conditions. Art and Lekeux (1988) as well as Honstein (1977) observed no changes in the breathing pattern of ponies and horses with changes in ambient temperature and humidity. On the contrary Morgan (1997a) found increased respiration rates for horses exposed to temperatures above 20 °C for one hour. Ousey (1992) described increasing respiration quotients for foals with decreasing temperatures and Cymbaluk and Christison (1993) showed a reduction of 27% in the respiration rate in yearlings under cold environment exposure.

Blood parameters

Different blood parameters exhibit changes at extreme ambient temperatures that can be attributed to thermoregulation. However, only little information regarding changes in blood parameters of horses exposed to high and low ambient temperatures is available.

The thyroid gland produces the hormones thyroxin and triiodthyronine which are responsible for the basal metabolic activity. When these hormones are increasingly produced and released to the blood, the metabolic rate increases (Scheunert and Trautmann, 1987) Thus, the metabolic intensity can be determined by their concentration in the blood (Mejdell and Boe, 2005). High and low ambient temperatures lead to changes in the thyroid gland activity. With high ambient temperature less thyroxin is released to the blood to reduce the metabolic heat production. A cold environment leads to an increased thyroid gland activity with increased thyroxin concentrations in the blood and an increased heat production (Thompson, 1973; Penzlin, 1991). Irvine (1967) showed an increase in the thyroxin secretion rate of horses during adaptation and housing in a cold environment. In addition, Floris (1990) found effects of the season on the triiodthyronine concentration. Despite this, Mejdell and Boe (2005) could not detect an effect of ambient temperature on serum thyroxin concentration. Equal findings were reported by McBride (1985) for thyroxin and triiodthyronine. According to Sjaastad (2003) changes in the thyroid hormones secretion do not occur instantly after acute cold exposure since the thyroid gland needs to increase size before a higher secretion level is reached. Therefore, prolonged cold exposure is needed to increase thyroid hormone release.

During cold winter weather the energy available by feed consumption may not be sufficient and body fat is mobilized to cover the energy demand. Blood lipid concentrations can therefore change with temperature changes as monitored by Floris (1990). Likewise Ott (2005) reported an altered rate of glucose and lipid metabolism.

High environmental temperatures led to higher concentrations of total protein and urea nitrogen as well as adrenalin. The total protein and urea nitrogen concentrations decreased under low ambient temperatures (Floris et al., 1990; Ott, 2005).

Sweating

The heat loss via evaporation is the most effective heat loss mechanism under hot temperatures for the horse (Morgan et al., 1997). It has been suggested that when temperatures and humidity is high, evaporative heat loss is difficult (McCutcheon and Geor, 1996). At high ambient temperatures the amount of sweat lost by the horse increases (Ott, 2005). At 20 to 25 °C no or little cutaneous water vapour loss was recognized by Allen and Bligh (1969) while Johnson and Creed (1982) found a water vapour loss of $61 - 94$ g/m^2 per hour at these temperatures. At temperatures of 45 °C the losses were distinctly higher with $306 - 1210$ g/m^2 per hour. Though, fluctuating rates in sweat output of the horse were observed with increasing temperatures (Allen and Bligh, 1969; Johnson and Creed, 1982), whereas initial sweating occurs as pulses with a frequency of one pulse per minute, followed by a continuous emission of sweat (Johnson and Creed, 1982).

The sweat of the horse is unusual as it contains high concentrations of glycoproteins that reduces the surface tension and helps the sweat spread along the hair. Thus, a greater surface is formed helping to increase the evaporation and heat loss (Marlin, 2008; Kennedy, 2011). Furthermore, the proteins lead to foam formation preventing sweat from dripping down, because every lost drop of sweat is lost for thermoregulation. Additionally the foam formation increases the reflection of solar radiation.

With decreasing ambient temperatures the evaporative heat loss looses importance and other ways of heat dissipation become significant (Morgan et al., 1997).

The described changes in the physiological parameters due to exposure to heat usually decline after a few days since the animal acclimatize to the higher temperatures (Ott, 2005), e.g. by a reduction of the metabolic heat production and feed intake (Bianca, 1977).

Scope of thesis

The number of robust horses held in year round extensive outdoor housing systems increases continuously, thus exposing the animals to extreme environmental conditions. Wild horses are known to adapt to changing environmental conditions, e.g. by the accumulation of body fat, changes in activity or even a reduction in metabolic rate. It is the question if domesticated horses have the ability to adapt to the extreme climatic conditions as they are usually selected for high performance, short hair length and a less compact body stature. Extensive outdoor housing of horses is also associated with a restricted control over animals, especially in winter. Particular the body condition of ponies is difficult to assess and can result in inadequate feed supply that may cause health problems in the animals. This impairment of the animals' health has financial consequences caused by the therapy as well as ethological consequences by the impaired welfare.

Ten Shetland ponies were housed under semi-extensive conditions over a one year period. In summer the animals were kept on pasture with access to shelter, in winter the ponies were housed on a paddock with an adjacent stable. The effect of possible feed shortage in extensive horse keeping systems on animal health was simulated by feed restriction of five animals in winter.

The aim of this thesis was to investigate seasonal adaptive changes in locomotor activity, resting behaviour, physiological and blood parameters as well as in total body water and total water intake. The changes in the parameters can be used to draw conclusions on the adaptation capacity of these animals and to evaluate if the year round semi-extensive housing meets welfare requirements of the horses.

The more detailed aims of the study were:
- To determine the diurnal and circadian rhythm of activity and resting behaviour and to observe changes in the adaptation capacity of metabolic rate, body temperature regulation and resting heart rate in horses kept under semi-extensive conditions (Chapter 2).
- To observe whether the recorded data suggests the existence of hypometabolism in larger domestic mammals such as the horse (Chapter 2).
- To estimate seasonal adaptive changes of hormones and metabolites in the horse under semi-extensive conditions and during feed restriction (Chapter 3).

- To compile equilibration times measured for the isotopes Deuterium, Tritium and ^{18}O in animal studies and to identify impacts upon it (Chapter 4).

- To identify factors influencing the total water intake and the total body water in horses under semi-extensive housing conditions and to evaluate a prediction equation for total body water based on the isotope dilution technique (Chapter 5).

- To compare the adaptation mechanisms in the horse found in the present studies with strategies found in other domestic herbivores such as sheep, goats and cattle (Chapter 6).

- To evaluate if the semi-extensive outdoor housing leads to impairment of the animal welfare or if this housing system can be assessed as adequate for horses (Chapter 6).

References

Acworth, N. R. J. (2003). The healthy neonatal foal: routine examinations and preventative medicine. *Equine Vet. Educ.* 15, 207-211.

Alexander, G., J. W. Bennett and R. T. Gemmell (1975). Brown adipose-tissue in newborn calf *(Bos-Taurus)*. *J. Physiol. London* 244, 223-234.

Allen, T. E. and J. Bligh (1969). A comparative study of temporal patterns of cutaneous water vapour loss from some domesticated mammals with epitrichial sweat glands. *Comp. Biochem. Physiol.* 31, 347-363.

Arnold, G. W. and M. L. Dudzinski (1978). Ethology of Free-Ranging Domestic Animals. Elsevier Scientific Publishing Company, Amsterdam, Netherlands.

Arnold, W., T. Ruf and R. Kuntz (2006). Seasonal adjustment of energy budget in a large wild mammal, the Przewalski horse *(Equus ferus przewalskii)* II. Energy expenditure. *J. Exp. Biol.* 209, 4566-4573.

Arnold, W., T. Ruf, S. Reimoser, F. Tataruch, K. Onderscheka and F. Schober (2004). Nocturnal hypometabolism as an overwintering strategy of red deer *(Cervus elaphus)*. *Am. J. Physiol.-Reg. I.* 286, R174-R181.

Art, T. and P. Lekeux (1988). Effect of environmental-temperature and relative-humidity on breathing pattern and heart-rate in ponies during and after standardized exercise. *Vet. Rec.* 123, 295-299.

Autio, E. (2008). Loose Housing of Horses in a Cold Climate. Doctoral dissertation, University of Kuopio, Kuopio, Finland.

Berger, A., K. M. Scheibe, K. Eichhorn, A. Scheibe and J. Streich (1999). Diurnal and ultradian rhythms of behaviour in a mare group of Przewalski horse *(Equus ferus przewalskii)*, measured through one year under semi-reserve conditions. *Appl. Anim. Behav. Sci.* 64, 1-17.

Berger, A., K. M. Scheibe, K. Wollenweber, B. Patan, P. Schnitker, C. Herrman and K. D. Budras (2006). Jahresrhythmik von Aktivität, Nahrungsaufnahme, Lebendmasse und Hufentwicklung bei Wild- und Hauspferden in naturnahen Lebensbedingungen. In: Aktuelle Arbeiten zur artgemäßen Tierhaltung 2006. KTBL Schrift 448. KTBL, Darmstadt, pp 137-146.

Bianca, W. (1968). Neuzeitliche Erkenntnisse und Angaben der Bioklimatologie bei Haustieren. *Der Tierzüchter* 20, 438-442.

Bianca, W. (1971). Die Akklimatisation von Haustieren. *Der Tierzüchter* 23, 187-189.

Bianca, W. (1977). Thermoregulation durch Verhaltensweisen bei Haustieren. *Der Tierzüchter* 3, 109-113.

Bianca, W. (1979). Nutztier und Klima. *Der Tierzüchter* 31, 188-192.

Bligh, J. (1998). Mammalian homeothermy: An integrative thesis. *J. Therm. Biol.* 23, 143-258.

Bogner, H. and A. Grauvogl (1984). Verhalten landwirtschaftlicher Nutztiere. Eugen Ulmer Stuttgart, Germany.

Boyd, L. and K. A. Houpt (1994). Przewalski´s horse: The History and Biology of an Endangered Species. State University of New York Press, Albany, USA.

Cannon, B. and J. Nedergaard (2004). Brown adipose tissue: Function and physiological significance. *Physiol. Rev.* 84, 277-359.

Clark, J. A. (1994). Stables. In: Livestock Housing (eds C. M. Wathes and D. R. Charles), Cambridge University Press, Cambridge, UK, pp 379-403

Clarke, A. F. (1987). A review of environmental and host factors in relation to equine respiratory-disease. *Equine Vet. J.* 19, 435-441.

Crompton, A. W., C. R. Taylor and J. A. Jagger (1978). Evolution of homeothermy in mammals. *Nature* 272, 333-336.

Crowell-Davis, S. L. (1994). Daytime rest behavior of the Welsh pony *(Equus-Caballus)* mare and foal. *Appl. Anim. Behav. Sci.* 40, 197-210.

Cymbaluk, N. F. (1990). Cold housing effects on growth and nutrient demand of young horses. *J. Anim. Sci.* 68, 3152-3162.

Cymbaluk, N. F. and G. I. Christison (1989). Effects of diet and climate on growing horses. *J. Anim. Sci.* 67, 48-59.

Cymbaluk, N. F. and G. I. Christison (1993). Cold weather - does it affect foal growth. Fourth Int. Livestock Env. Symp., Am. Soc. Agric. Eng, St. Joseph, pp 23-30.

Davenport, J. (1992). Animal Life at Low Temperatures. Chapman & Hall, London, UK.

Dawkins, M. J. R. and D. Hull (1964). Brown adipose tissue and response of new-born rabbits to cold. *J. Physiol. London* 172, 216-238.

Duncan, P. (1980). Time-budgets of Camargue horses. 2. Time-budgets of adult horses and weaned sub-adults. *Behaviour* 72, 26-49.

Duncan, P. (1985). Time-budgets of Camargue horses. 3. Environmental-influences. *Behaviour* 92, 188-208.

Else, P. L. and A. J. Hulbert (1981). Comparison of the mammal machine and the reptile machine - energy-production. *Am. J. Physiol.* 240, R3-R9.

Floris, B., P. P. Bini, P. Nuvole and F. G. Di Meglio (1990). The horses of Giara: variation of the thyroid activity and certain blood parameters between winter and summer. *Boll. Soc. Ital. Biol. Sper.* 66, 849-56.

Foster, D. O. and M. L. Frydman (1978). Nonshivering thermogenesis in rat .2. Measurements of blood-flow with microspheres point to brown adipose-tissue as dominant site of calorigenesis induced by noradrenaline. *Can. J. Physiol. Pharm.* 56, 110-122.

French, A. R. (1982). Effects of temperature on the duration of arousal episodes during hibernation. *J. Appl. Physiol.* 52, 216-220.

Gattermann, R. (2006). Wörterbuch der Verhaltensbiologie der Tiere und des Menschen. Spektrum Akademischer Verlag, Heidelberg.

Geiser, F. (2004). Metabolic rate and body temperature reduction during hibernation and daily torpor. *Annu. Rev. Physiol.* 66, 239-274.

Geiser, F. and T. Ruf (1995). Hibernation versus daily torpor in mammals and birds - physiological variables and classification of torpor patterns. *Physiol. Zool.* 68, 935-966.

Heldmaier, G., S. Ortmann and R. Elvert (2004). Natural hypometabolism during hibernation and daily torpor in mammals. *Resp. Physiol. Neurobi.* 141, 317-329.

Hensel, H. (1966). Allgemeine Sinnesphysiologie: Hautsinne, Geschmack, Geruch. Springer Verlag, Berlin, Germany.

Hetem, R. S., S. K. Maloney, A. Fuller, L. C. R. Meyer and D. Mitchell (2007). Validation of a biotelemetric technique, using ambulatory miniature black globe thermometers, to quantify thermoregulatory behaviour in ungulates. *J. Exp. Zool. Part A* 307, 342-356.

Hines, M. T. (2004). Changes in Body Temperature. In: Equine Internal Medicine (eds S. M. Reed, W. M. Bayly and D. C. Sellon), Saunders Elsevier, St Louis, USA, pp 148-155.

Honstein, R. N. and D. E. Monty (1977). Physiologic responses of horse to a hot, arid environment. *Am. J. Vet. Res.* 38, 1041-1043.

Houpt, K. A. (2005). Maintenance Behaviours. In: The Domestic Horse (eds D. S. Mills and S. M. McDonnell), Cambridge University Press, Cambridge, UK, pp 94–109.

Irvine, C. H. G. (1967). Thyroxine secretion rate in horse in various physiological states. *J. Endocrinol.* 39, 313-320.

Johnson, K. G. and K. E. Creed (1982). Sweating in the intact horse and isolated perfused horse skin. *Comp. Biochem. Phys. C* 73, 259-264.

Kayser, C. (1961). The Physiology of Natural Hibernation. Pergamon Press, New York, USA.

Kennedy, M. W. (2011). Latherin and other biocompatible surfactant proteins. *Biochem. Soc. T.* 39, 1017-1022.

Knickel, U. R., C. Wilczeck and K. Jöst (2002). MemoVet - Praxis-Leitfaden Tiermedizin. Schattauer, Stuttgart, Germany.

Kolb, E. (1967). Lehrbuch der Physiologie der Haustiere. Gustav Fischer Verlag, Jena, Germany.

Kooistra, L. H. and O. J. Ginther (1975). Effect of photoperiod on reproductive activity and hair in mares. *Am. J. Vet. Res.* 36, 1413-1419.

Kortner, G. and F. Geiser (2000a). The temporal organization of daily torpor and hibernation: Circadian and circannual rhythms. *Chronobiol. Int.* 17, 103-128.

Kortner, G. and F. Geiser (2000b). Torpor and activity patterns in free-ranging sugar gliders *Petaurus breviceps (Marsupialia). Oecologia* 123, 350-357.

Langlois, B. (1994). Inter-breed variation in the horse with regard to cold adaptation - a review. *Livest. Prod. Sci.* 40, 1-7.

MacMillen, R. E. (1965). Aestivation in cactus mouse *Peromyscus eremicus. Comp. Biochem. Phys.* 16, 227-248.

Marlin, D. (2008). Thermoregulation in the Horse at Rest and During Exercise. In Nutrition of the Exercising Horse (eds M. Saastamoinen and W. Martin-Rosset), Wageningen Academic Publishers, Wageningen, Netherlands, pp 71-83.

McArthur, A. J. (1991). Metabolism of homeotherms in the cold and estimation of thermal insulation. *J. Therm. Biol.* 16, 149-155.

McBride, G. E., R. J. Christopherson and W. Sauer (1985). Metabolic-rate and plasma thyroid-hormone concentrations of mature horses in response to changes in ambient-temperature. *Can. J. Anim. Sci.* 65, 375-382.

McConaghy, F. F., D. R. Hodgson, R. J. Rose and J. R. S. Hales (1996). Redistribution of cardiac output in response to heat exposure in the pony. *Equine Vet. J.* 28, 42–46.

McCutcheon, L. J. and R. J. Geor (1996). Sweat fluid and ion losses in horses during training and competition in cool vs. hot ambient conditions: implications for ion supplementation. *Equine Vet. J. Supplement*, 54-62.

Mejdell, C. M. and K. E. Boe (2005). Responses to climatic variables of horses housed outdoors under Nordic winter conditions. *Can. J. Anim. Sci.* 85, 301-308.

Mills, D. and McDonnell (2005). The Domestic Horse. Cambridge University Press, Cambridge, UK.

Mogg, K. C. and C. C. Pollitt (1992). Hoof and distal limb surface-temperature in the normal pony under constant and changing ambient-temperatures. *Equine Vet. J.* 24, 134-139.

Morgan, K. (1996). Short-term thermoregulatory responses of horses to brief changes in ambient temperatures. Doctoral dissertation, Swedish University of Agricultural Sciences, Uppsala, Sweden.

Morgan, K. (1997a). Effects of short-term changes in ambient air temperature or altered insulation in horses. *J. Therm. Biol.* 22, 187-194.

Morgan, K. (1997b). Thermal insurance of peripheral tissue and coat in sport horses. *J. Therm. Biol.* 22, 169-175.

Morgan, K., A. Ehrlemark and K. Sallvik (1997). Dissipation of heat from standing horses exposed to ambient temperatures between -3 degrees C and 37 degrees C. *J. Therm. Biol.* 22, 177-186.

Morhardt, J. E. (1970). Body temperatures of white-footed mice *(Peromyscus Sp.)* during daily torpor. *Comp. Biochem. Physiol.* 33, 423-457.

Ott, E. (2005). Influence of temperature stress on the energy and protein metabolism and requirements of the working horse. *Livest. Prod. Sci.* 92, 123-130.

Ousey, J. C., A. J. McArthur, P. R. Murgatroyd, J. H. Stewart and P. D. Rossdale (1992). Thermoregulation and total-body insulation in the neonatal foal. *J. Therm. Biol.* 17, 1-10.

Palmer, S. E. (1983). Effect of ambient-temperature upon the surface-temperature of the equine limb. *Am. J. Vet. Res.* 44, 1098-1101.

Penzlin, H. (1991). Lehrbuch der Tierphysiologie. Gustav Fischer Verlag, Jena, Germany.

Piccione, G., G. Caola and R. Refinetti (2002). The circadian rhythm of body temperature of the horse. *Biol. Rhythm Res.* 33, 113-119.

Radostits, O. M., C. C. Gay, D. C. Blood and K. W. Hinchcliff (2005). Veterinary Medicine - A Textbook of the Diseases of Cattle, Sheep, Pigs and Horses. 9th Edition. W.B. Sauders, Philadelphia, USA.

Renecker, L. A. and R. J. Hudson (1985). Telemetered heart-rate as an index of energy-expenditure in moose *(Alces-alces)*. *Comp. Biochem. Phys. A* 82, 161-165.

Scheibe, K. M. and W. J. Streich (2003). Annual rhythm of body weight in Przewalski horses *(Equus ferus przewalskii)*. *Biol. Rhythm Res.* 34, 383-395.

Scheunert, A. and A. Trautmann (1987). Lehrbuch der Veterinär-Physiologie. 7th Edition. Verlag Paul Parey, Berlin Hamburg.

Schmidt-Nielsen, K. (1997). Animal Physiology - Adaptation and Environment. 5th Edition. Cambridge University Press, Cambridge, UK.

Schrader, L. (2009). Tierschutz und Tierhaltung in der Milchviehhaltung. *Züchtungskunde* 81, 414 - 420.

Signer, C., T. Ruf and W. Arnold (2011). Hypometabolism and basking: the strategies of Alpine ibex to endure harsh over-wintering conditions. *Funct. Ecol.* 25, 537-547.

Singer, D. (2007). Why 37 degrees C? Evolutionary fundamentals of thermoregulation. *Anaesthesist* 56, 899-906.

Sjaastad, O. V., K. Hove and O. Sand (2003). Phsiology of Domestic Animals. Scandinavian Veterinary Press, Oslo, Norway.

Symonds, M. E., M. J. Bryant, L. Clarke, C. J. Darby and M. A. Lomax (1992). Effect of maternal cold-exposure on brown adipose-tissue and thermogenesis in the neonatal lamb. *J. Physiol. London* 455, 487-502.

Thompson, G. E. (1973). Review of progress of dairy science - climatic physiology of cattle. *J. Dairy Res.* 40, 441-473.

Twente, J. W., J. Twente and R. M. Moy (1977). Regulation of arousal from hibernation by temperature in 3 species of citellus. *J. Appl. Physiol.* 42, 191-195.

Vogel, C. (2006). The Complete Performance Horse. David and Charles, Cincinnati, USA.

von Engelhardt, W. and G. Breves (2000). Physiologie der Haustiere. 3rd Edition. Enke Verlag, Stuttgart, Germany.

Wittke, G. (1972). Physiologie der Haustiere: kurzes Lehrbuch für Studierende der Agrarwissenschaften, Veterinärmedizin und Biologie. Paul Parey Verlag, Berlin, Germany.

Wollenweber, K. (2007). Das Verhalten einer Pferdeherde (Liebentaler Pferde) unter naturbelassenen Lebensbedingungen im Hinblick auf chronobiologische Aspekte, klimatische Einflüsse sowie deren Raumnutzung, Doctoral Dissertation. FU Berlin, Berlin, Germany.

Wood, C. and A. Griffin (2009). Equine thermoregulation. http://www.extension.org/pages/11557/equine-thermoregulation, 15.11.2011

Young, B. A. and J. Coote (1973). Some effects of cold on horses. Feeder's Day Report, University of Alberta, Edmonton, Canada.

Zeeb, K. (1994). Möglichkeiten der ganzjährigen Freilandhaltung von Pferden. *Deut. Tierarztl. Woch.* 101, 122-123.

Zeitler-Feicht, M. (2001). Handbuch Pferdeverhalten. Ursache, Therapie und Prophylaxe. Ulmer-Verlag, Stuttgart, Germany.

CHAPTER 2

Adaptation strategies to seasonal changes in environmental conditions of a domesticated horse breed, the Shetland pony *(Equus ferus caballus)*

L. Brinkmann, M. Gerken, A. Riek

Department of Animals Sciences, University of Göttingen, Albrecht-Thaer Weg 3, 37075 Göttingen, Germany

Journal of Experimental Biology (2011) 215: 1061-1068

Submitted on August 25, 2011

Accepted on December 5, 2011

doi: 10.1242/jeb.064832

This article has been reproduced with permission from the Journal of Experimental Biology

Adaptation strategies to seasonal changes in environmental conditions of a domesticated horse breed, the Shetland pony (*Equus ferus caballus*)

Lea Brinkmann, Martina Gerken and Alexander Riek

Department of Animal Sciences, University of Göttingen, Albrecht-Thaer-Weg 3, 37075 Göttingen, Germany

Summary

Recent results suggest that the wild ancestor of the horse, the Przewalski horse, exhibits signs of a hypometabolism. However, there are speculations that domestic animals lost the ability to reduce energy expenditure during food shortage and adverse environmental conditions. Therefore, we investigated physiological and behavioural strategies employed by a robust domesticated horse breed, the Shetland pony, over the course of a year under temperate conditions by measuring ambient temperature (T_a), subcutaneous temperature (T_s), locomotor activity (LA), lying time, resting heart rate, body mass and the body condition score. Ten animals were kept on pasture in summer and in open stables in winter; further, in winter the animals were allocated into one control and one feed-restricted group of 5 animals each to simulate natural seasonal food shortage. The annual course of the mean daily T_s of all horses showed distinct fluctuations from a mean of 35.6 ± 0.5 °C, with higher variations in summer than in winter. Diurnal amplitudes in T_s were highest ($P < 0.001$) in April (12.6 °C) and lowest in January (4.0 °C), with a nadir around dawn and a peak around mid-day. The feed-restricted group had a significantly lower daily T_s compared with the control group on cold winter days, with T_a values below 0 °C. Mean annual heart rate and LA followed T_a closely. Heart rate of the feed-restricted animals significantly decreased from a mean of 52.8 ± 8.1 beats min^{-1} in summer to 29 ± 3.9 beats min^{-1} in winter and differed from the control group ($P < 0.001$). Mean daily LA was lowest at the end of winter (7,000 activity impulses day^{-1}) and highest in summer (25,000 activity impulses day^{-1}). Our results show that Shetland ponies exhibit signs of a winter hypometabolism indicated by reduced heart rate and T_s. Thus, domesticated horses seem to have maintained the capacity for seasonal adaptation to environmental conditions by seasonal fluctuations in their metabolic rate.

Key words: Shetland pony, hypometabolism, body temperature, metabolic rate, feed restriction, locomotion.

Introduction

Endothermic animals, like the horse, usually keep their body temperature (T_b) within a narrow limit with changing environmental conditions (Schmidt-Nielsen, 1997; Singer, 2007). However, this comes at a high energetic cost. Endothermic animals can therefore face a twofold challenge. In harsh environmental conditions, the availability and quality of food is limited but the energy requirement to maintain T_b is high (Arnold et al., 2006). Besides a small degree of a daily reduction in metabolism during the 24h cycle of rest and activity regularly observed in larger mammals (Langman and Maloiy, 1989; Taylor and Lyman, 1967), small animals may enter intense forms of hypometabolism, such as hibernation, prolonged torpor, or daily torpor (Geiser, 1988; Heldmaier et al., 1989). These adaptive mechanisms, employed to save energy, include the reduction of the metabolic rate (MR), a decrease of T_b, and a reduced heart and breathing rate (Heldmaier et al., 2004; Singer, 2007). Large animals, with the exception of bears, normally lack the ability to enter torpor (Arnold et al., 2006), which is generally restricted to animals weighing less than 1000 g (Geiser and Ruf, 1995). Nevertheless, there are indications that some ungulates can exhibit some form of an energy-saving mechanism when ambient temperatures (T_a) are low and little food is available. This phenomenon has been shown for large ungulates such as roe deer (Weiner, 1977), red deer (Arnold et al., 2004) and recently also for the wild ancestor of the horse, the Przewalski horse (*Equus ferus przewalski*) (Arnold et al., 2006).

There are speculations that livestock species lost the ability to reduce their energy expenditure under food shortages and adverse environmental conditions, because they were housed and selected under constant nutrient supply without selection pressure for maintaining a seasonal, cyclic adaptation (Price, 1984). However, empirical confirmations of a supposedly lower adaptability of livestock species compared with their wild counterparts are still missing. Therefore, we investigated whether the physiological capabilities of seasonal adjustments are still present in a livestock species when kept under long-term semi-natural outdoor conditions. We chose Shetland ponies (*Equus ferus caballus*, Linnaeus 1758), a robust horse breed that, according to the domestication model of Lauvergne (Lauvergne, 1994), can be characterized as a primary population. We studied whether there exist seasonal fluctuations in crucial physiological and behavioural parameters such as locomotor activity (LA), resting time, subcutaneous temperature (T_s) and heart rate similar to those described for wild horses (Kuntz et al., 2006; Arnold et al., 2006). Furthermore, it is unclear whether a suggested form of hypometabolism in wild horses (Arnold et al., 2006) as an overwintering strategy is still present in domestic horses. Therefore, we simulated the food shortage found in natural habitats in winter by reducing the food supply to 70% of the required energy demand for one group in the winter months while the other group had permanent access to food. The results of our study will provide more in-depth knowledge on the adaptive physiological capabilities of a domesticated species. Hence, the results of our study will also contribute to detection of possible modifications in the adaption ability of domesticated horses. Therefore, we tested the hypothesis that Shetland ponies, a robust horse breed that can be characterised as a primary population after the first steps of domestication, still exhibit signs of a hypometabolism similar to those described for its wild ancestor, the Przewalski horse.

Materials and methods

Animals and management

The study was undertaken at the Department of Animal Sciences at the University of Göttingen, Germany, and lasted for 1 year from February 2010 to February 2011. Initially, we started with eight Shetland pony mares; however, the herd was increased to 10 mares from the beginning of May 2010. The age of the animals ranged between 4 and 16 years. Animals were raised on different farms in Germany and The Netherlands and were accustomed to outdoor housing systems.

During our study, animals were kept either on paddock or on pasture. From February 2010 until the end of May 2010, ponies were housed in two groups of five ponies each in two identical paddocks. Each paddock measured 210 m^2 and had a permanent access to a pen measuring 6.40 x 2.95 m that was covered with straw. Two large exits allowed the animals to enter and leave without rank conflicts. Both pens were equipped with five feeding stands each (1.35 x 1.60 x 0.55 m, height x length x width) to ensure that every horse was able to ingest the required food. The stable was not heated and hence T_a values inside and outside the stable were comparable. The light-dark cycle inside the stable fluctuated according to the natural photoperiod. From February 2010 until the end of May 2010, all ponies received the same feeding treatment with straw *ad libitum* and 5 kg hay per 100 kg body mass per day. Additionally, ponies received 22 g mineral mixture per 100 kg body mass per day (Derby® Mineralpellets, Derby Spezialfutter GmbH, Münster, Germany) and 580 g concentrate per 100 kg body mass per day (Derby® Standard, Derby Spezialfutter GmbH). Water was available *ad libitum* throughout the experiment at a frost-proof watering place. From the end of May 2010 until the middle of October 2010, all animals were kept on permanent pastures with access to two 4.0 x 3.6 m shelters (plastic tents). In addition to grass, ponies were offered a small amount of hay, straw and mineral supplement. From the middle of October 2010 until the end of our study (February 2011), animals were kept on the same paddocks described above. During that time, ponies were allocated according to their body mass and body condition score (BCS) into a control group (CG) and a treatment group (TG) of five animals each, resulting in a comparable mean BCS for both groups. In the first 2 weeks, both groups were fed identically as described above. However, although from the beginning of November 2010 the control group was fed as before, the treatment group was fed restrictively to simulate the limited availability of food in autumn and winter under natural conditions. The amount of food offered in the TG animals was gradually reduced from 100% to 80% of the recommended energy and protein requirements for Shetland ponies (Gesellschaft für Ernährungsphysiologie, 1994) until 21 January and to a further 70% until 28 February.

The research conducted in this study was performed in accordance with the current laws regulating animal welfare and experiments with animals in Germany.

Body mass, body condition score, ambient temperature and hair length

Body mass for all animals was recorded on a bi-weekly basis throughout the course of the study (February 2010 – February 2011) with a mobile scale (Weighing System MP 800, resolution: 0.1 kg, Patura KG, Laudenbach, Germany). The BCS, a palpable and visual assessment of the degree of fatness in the neck, back, ribs and pelvis, was taken monthly after Carroll and Huntington (Carroll and Huntington, 1988). T_a and relative humidity (RH) were recorded continuously throughout the study with miniature data loggers at 10 minutes intervals (Tiny view TV 1500, temperature resolution: 0.25 °C, humidity resolution: 0.5%, Gemini, Chichester, West Sussex, UK). Hair length for each pony was measured once in summer (July 2010) and once in winter (February 2011) with a cardboard template (7 x 10.5 cm) that was inserted into the hair coat of the flank, determining the length of guard and undercoat hairs.

Locomotor activity

The LA for each pony was measured continuously for the entire study period using activity-lying time-temperature-pedometers (ALT-Pedometer, Engineering Office Holz, Falkenhagen, Germany) tied to the foreleg above the pastern. Pedometers were lined with a silicon pad to avoid pressure marks on the foreleg. The pedometer (125 g mass; 6 x 5 x 2 cm, length x width x height) consisted of four acceleration sensors, allowing the measurement of both locomotion and lying time. The LA was recorded as activity impulses generated by the front leg with a maximum resolution of two impulses per second. Furthermore, sensors detected the position (lateral or ventral) of the pedometer every 15 s, thus allowing us to determine the total time spent lying. The recorded data were added up to 15 minutes intervals and saved to an on-board storage device.

Subcutaneous temperature

For the continuous recording of T_s, a high-resolution, real-time synchronised miniature temperature data logger (i-Button DS1922L-F5, resolution: 0.5 °C, Maxim Integrated Products, Sunnyvale, CA, USA) was implanted subcutaneously on the right side of each animal's neck. The implants (3.3 g mass; 1.74 x 1.74 x 0.6 cm) were coated with a medical silicon layer to waterproof them and prevent any inflammatory response. Prior to the implantation of the loggers, a 10 x 10 cm area at the right side of the animal's neck was shaved, washed and soaked with iodine solution. A local anaesthetic (2% xylocain with adrenalin, Rompun, Bayer Animal Health, Leverkusen, Germany) was administered subcutaneously 10 minutes before the surgery. Subsequently, a 4 cm vertical incision was performed resulting in a subcutaneous pocket (4 x 4 cm) in which the logger was positioned. The incision was then closed using a skin stapler (Weck Visistat 35 W, Teleflex Medical Europe Ltd., Athlone, Ireland). All animals received a broad-spectrum antibiotic (Procain-Penicillin G, Pfizer AG, New York City, NY, USA) for the next six days. The staples were removed after 12 days. The removal of the implants after 12 months was performed the same way as described above.

Each logger was programmed to measure and record T_s every 2h for the 12 month period, resulting in 4488 records for each animal. Clock time of the loggers was synchronised across ponies.

Resting heart rate and rectal temperature

The heart rate was recorded bi-weekly with a stethoscope between 12:00 and 13:00 h, during which heart rate was determined three times for 60s. Before the measurements, animals were at rest for at least 5 minutes and were loosely tied with a rope but had visual contact with their conspecifics. The ponies were used to the handling and thus any impact of the measuring procedure on the heart rate recordings was assumed to be minimal. Rectal temperature (T_r) was recorded bi-weekly for all experimental animals during the feeding trial from November 2010 to February 2011 using a clinical thermometer (resolution: 0.1 °C, AEG, Nürnberg, Germany) with the probe inserted approximately 5 cm into the rectum.

Statistical Analysis

All raw data recorded with the pedometers were checked for obviously false records. In a few instances, pedometers became detached from the animals` legs for a maximum of two days and these records were discarded from statistical analyses. Two-hourly, daily and monthly averages for T_s, LA and lying duration were calculated for each animal. Data for heart rate, LA, lying duration, T_s and T_r were then subjected to an ANOVA. We used mixed modelling (PROC MIXED) to control for repeated measurements from the same individuals. To model correlation within the two feeding groups across time and between the two feeding groups, we first determined the appropriate covariance structure for the dataset based on Akaike's Information Criterion adjusted for small sizes (AICc) values. Differences between the two feeding groups were analysed using data recorded from mid-November 2010 until the end of February 2011 only. Furthermore, individual averages for T_s, T_r, LA, lying duration, heart rate, BCS and body mass were calculated for three periods (i.e. first winter: February 2010 – May 2010; summer: May 2010 – October 2010; second winter: October 2010 – February 2011) and averages for hair length were calculated across two measurement month and analysed for differences using the Mann-Whitney-*U*-test as an integrated *post hoc* test in the MIXED procedure. Data are expressed as means ± s.d., where *N* is the number of individuals and n is the number of measurements. Statistical analyses were performed with the program package SAS version 9.2 (SAS, 2008).

Results

Climate data, body mass and body condition score
The climatic conditions during the 12 months of the experiment were within the long-term climate recordings of the area. During the time of our study we experienced one hot summer month (July 2010; mean T_a=20.3 °C, maximum T_a=34.9 °C, minimum T_a=6.4 °C, mean RH=70.7%, maximum RH=100%, minimum RH=30%) and low T_a values in both winter periods (Fig. 1). Rainfall occurred on 129 of the 372 study days, with an annual rainfall of 632.8 mm. The second winter period included 44 days with snow heights over 1 cm and 84 days with ground frost.
Body mass and BCS of the ponies were relatively constant during the first winter period but increased during the summer period with peak values in September (Body mass: 158 ± 38 kg; BCS: 4.3 ± 0.6) and October (Body mass: 160.8 ± 38 kg; BCS: 4.6 ± 0.4). During the second winter period, the TG animals lost on average 20% of their initial body mass, with a significant drop from 162.4 ± 41 kg in Oct 2010 to 132 ± 36 kg in Feb 2011 (body mass (kg) = 165.7 - 4.18 month, R^2 =0.12, $P < 0.05$). A similar development was observed for the BCS, which decreased significantly from 4.6 ± 0.4 points in October 2010 to 2.4 ± 0.8 points in February 2011 (BCS = 5.4 -

0.73 month, R^2 = 0.65, P < 0.001) whereas mean body mass and BCS of the CG group did not change during this time (body mass (kg) = 149.8 + 0.22 month, R^2 = 0, P = 0.93; BCS = 3.4 + 0.09 month, R^2 = 0.03, P = 0.43). However, the two different feeding groups did not differ significantly in body mass and BCS because of large individual variations.

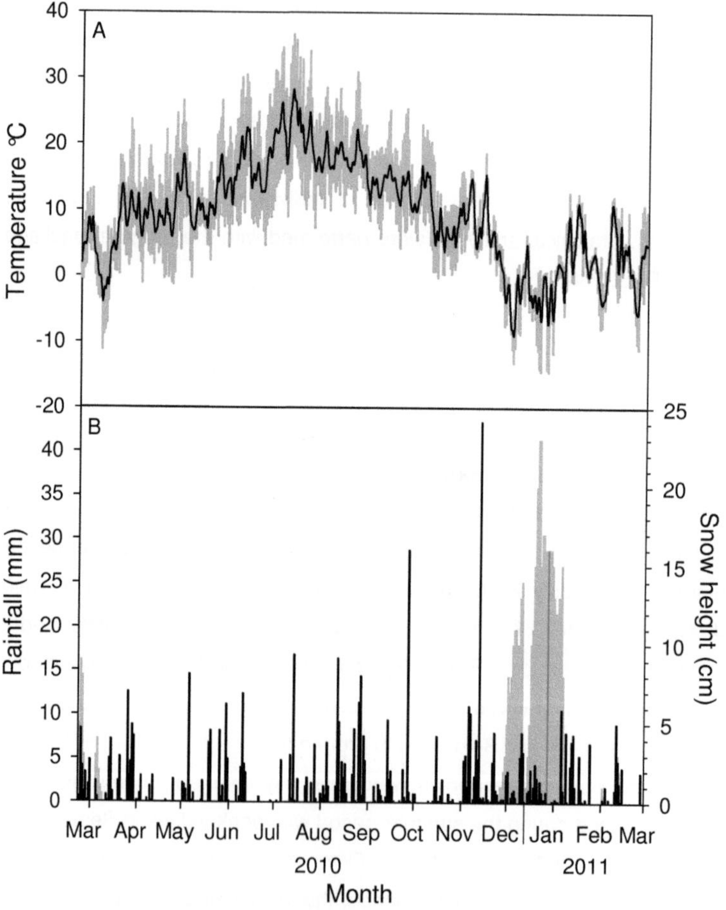

Fig. 1 Annual variation in ambient temperature (T_a) and precipitation at the Institute of Animal Sciences (University of Göttingen) over the course of the study. (A) Solid line, daily mean T_a; grey shading, daily maximum and minimum T_a. (B) Black columns, daily rainfall (mm); grey columns, snow height (cm).

Hair length differed significantly ($P < 0.001$) between summer (guard hair, 1.0 ± 0.3 cm; undercoat hair: 1.0 ± 0.3) and winter (guard hair, 6.3 ± 0.8 cm; undercoat hair, 4.1 ± 0.4 cm) for all animals. Although no differences ($P > 0.05$) were found in the undercoat hair length between the two feeding groups, guard hair length was significantly ($P < 0.05$) longer in the TG ponies compared with the CG ponies at the end of the study (February 2011: 6.7 ± 0.6 cm vs 5.9 ± 0.7 cm, respectively).

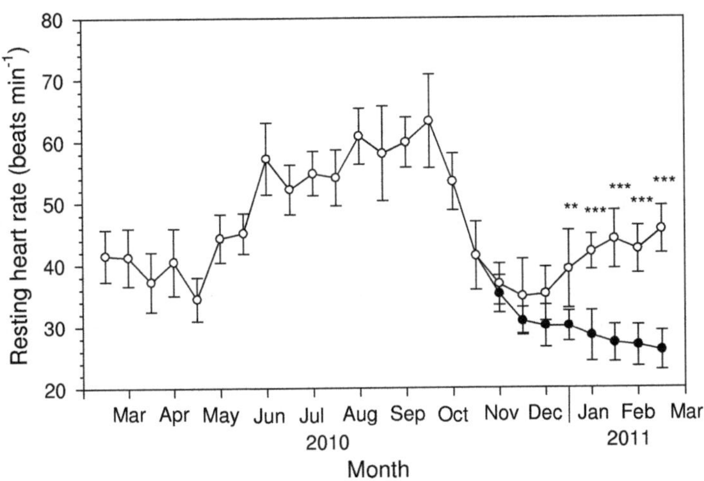

Fig. 2 Seasonal changes in resting heart rate (means ± s.d.) of eight (February - May 2010) and ten Shetland pony mares (June - October 2010). From November 2010, animals were divided into a feed-restricted group ($N = 5$, filled circles) and a control group ($N = 5$, open circles; see Material and methods for details). Significant differences between feeding groups are indicated by asterisks (** $P < 0.01$, *** $P < 0.001$).

Heart rate

Mean heart rate of all animals varied considerably throughout the seasons (Fig. 2) and the annual pattern followed the trend of both T_a (Fig. 1) and LA (Fig. 3B), with low values in the winter months and high values in the summer months, peaking in September (63.2 ± 7.6 beats min^{-1}). Correlations of the annual means between heart rate and LA ($r = 0.77$, $P < 0.001$) as well as between heart rate and T_a ($r = 0.70$, $P <$

0.001) were high and significant. The heart rate of the TG animals significantly decreased from their summer (May – October) mean of 52.8 ± 8.1 beats min^{-1} to 29 ± 3.9 beats min^{-1} in the second winter period (October – February). Furthermore, mean heart rate differed significantly ($P < 0.001$) between TG and CG animals in the second winter period (October – February, 29 ± 3.9 beats min^{-1} *vs* 40.3 ± 5.7 beats min^{-1}, Fig. 2). The highest difference in heart rate between TG and CG animals was observed at the end of our study in February 2011 (26.0 ± 3.2 beats min^{-1} *vs* 45.6 ± 3.8 beats min^{-1}, respectively; $P < 0.001$), when animals were fed restrictively for 4 months.

Subcutaneous and rectal temperature

The mean T_s over the entire study period was 35.6 ± 0.5 °C. The annual course of the mean daily T_s of all ponies showed distinct fluctuations. During spring and summer, T_s fluctuated considerably, as evidenced by high variations in mean minimum ($T_{s,min}$) and mean maximum T_s ($T_{s,max}$) (Fig. 3A). However, these high variations decreased significantly ($P < 0.001$) during autumn and winter. Daily $T_{s,min}$ was lowest in April (28.2 °C) and highest in October (35.7 °C). In contrast, daily $T_{s,max}$ was lowest in December (35.9 °C) and highest in June (38.2 °C). Daily fluctuations in T_s were significantly higher in summer than in winter ($P < 0.001$; Fig. 4). Furthermore, T_s was significantly lower in TG animals on 52 days of the 120 days of feed restriction compared with the CG animals, especially during cold days when T_a values were constantly below 0 °C ($P < 0.05$; Fig. 5). During the entire study period, the lowest daily mean T_s and $T_{s,min}$ typically appeared in the early morning hours around dawn, usually coinciding with the lowest daily T_a (Fig. 5). In the early daylight hours, T_s increased and stayed either fairly constant during cool or cold days or increased to a peak at around mid-day during warm or hot days (Fig. 4, 5)

The overall correlation between mean daily T_s and mean daily T_a was positive and moderate ($r = 0.38$, $P < 0.001$), whereas the correlation between $T_{s,max}$ and T_a was higher ($r = 0.64$, $P < 0.001$). In contrast, the correlation between $T_{s,min}$ and T_a was negative ($r = -0.20$, $P < 0.001$). The relationship between T_s and T_a in winter months (October - February) was best described by a linear regression, $T_s = 35.41 + 0.057$ T_a ($R^2 = 0.50$, $P < 0.001$), whereas in summer months (May - October) the relationship between T_a and T_s only loosely followed a power function, $T_s = 31.56\, T_a^{0.05}$ ($R^2 = 0.25$, $P < 0.01$).

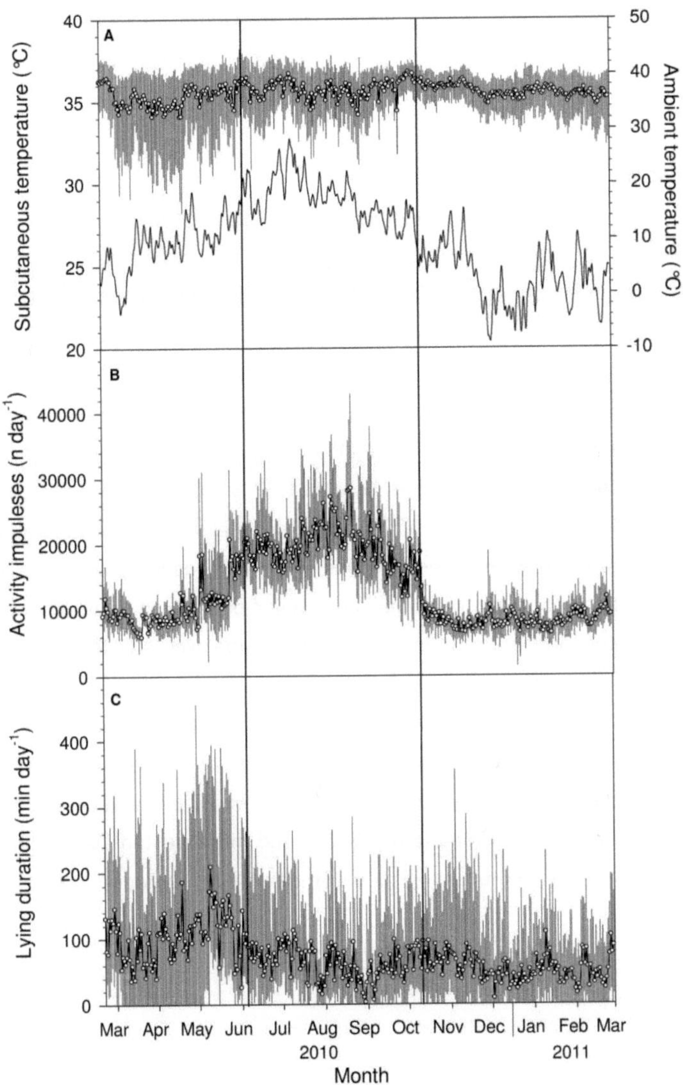

Fig. 3 Mean seasonal variation in (A) subcutaneous temperature (B) locomotor activity and (C) resting behaviour of Shetland pony mares (daily means with daily mean minimum and maximum). The solid line in A indicates the corresponding ambient temperature and the vertical lines spanning A-C indicate the changes from paddock housing (March - May 2010) to pasture (May – October 2010) and back to paddock housing (October 2010 – Feb ruary 2011).

During the feeding trial, the T_r in December, January and February was significantly ($P < 0.05$) lower in TG than in CG animals (TG: 36.8 ± 0.3, 36.7 ± 0.4 and 36.4 ± 0.3 °C; CG: 37.5 ± 0.3, 37.5 ± 0.3 and 37.5 ± 0.4 °C for December, January and February, respectively).

Locomotor activity and resting behaviour

The mean daily LA showed considerable variations throughout the course of the year (Fig. 3B), with the lowest values occurring during the winter month (7000 activity impulses day^{-1} in March) and the highest during the summer months (25,000 activity impulses day^{-1} in August). The mean LA values in summer (May – October) were significantly higher than those in winter (October – February, $P < 0.001$). However, no significant differences could be detected between both feeding groups ($P > 0.05$). LA followed a pattern similar to that of T_a (Fig. 3A, B), with a relatively high correlation between both parameters ($r = 0.64$, $P < 0.001$). During the diurnal LA rhythm, the highest numbers of activity impulses per hour were recorded in the morning around dawn, with a second peak frequently occurring around dusk. This pattern was observed in both summer and winter periods. In summer, the diurnal variation followed an approximately sinusoidal shape whereas the activity peaks were less distinct in winter (Fig. 4).

Daily lying times were generally low throughout the entire study period (70.1 ± 67.0 minute day^{-1}, $4.9 \pm 4.6\%$ of the day lying; Fig. 3C). Resting usually occurred at night, with the highest lying frequency observed before dawn. The highest mean daily lying durations were recorded in May ($8.3 \pm 6.4\%$) and the lowest in December ($2.72 \pm 3.1\%$). No significant differences ($P > 0.05$) were found between the two feeding groups.

Discussion

In our study we showed for the first time that domesticated horses decrease their T_s and heart rate, similar to that observed in their wild ancestor the Przewalski horse, when exposed to long-term semi-natural conditions including a harsh winter. These changes in T_s and heart rate can be interpreted as signs of a hypometabolism.

Seasonal changes in behaviour

Rhythms of physiological functions and their related behaviours are essential components of the relationship between the animals and their environment (Scheibe et al., 1999). Homeothermic animals attempt to keep their endogenous milieu on a stable level and exogenous disturbances can lead to changing behaviour as one strategy of the organism to overcome the interruption (McFarland, 1985). The LA can act as a final correcting element in a long chain of endogenous processes (Aschoff, 1962), and its circadian rhythm may give information about the adaptation ability of the animal in question (Scheibe et al., 1999). In our study, the LA of Shetland ponies under extensive housing (i.e. on pasture) was highly dependent on the T_a. Although we could not distinguish between activity due to grazing or foraging and other LAs, according to previous studies in horses (Duncan, 1985; Boyd and Bandi, 2002; Lamoot and Hoffmann, 2004) it is reasonable to assume that most of the LA recorded can be attributed to feeding behaviour. Even though horses were kept on paddock with access to an outdoor stable during winter in our study, their annual LA pattern agrees well with observations described for wild herbivores such as the Przewalski horse (Berger et al., 1999; Arnold et al., 2006; Berger et al., 2006) and roe deer (Berger et al., 2002). The diurnal LA rhythm in our study was related to the photoperiod and was in its pattern similar throughout much of the year, with higher activities during daylight than during the night (Fig. 4), which is in agreement with results for Przewalski horses (Boyd et al., 1988; Berger et al., 1999), cattle and sheep (Arnold, 1984). However, in winter, the LA during daylight differed less from that at night compared with during summer. Similarly, Berger (Berger et al.,1999) and Duncan (Duncan 1985) noted that horses were mostly inactive on cold winter days and restricted their activity to feeding as a mechanism of energy conservation (Malechek and Smith, 1976; Berger et al., 2006). We therefore assume that our horses saved energy in winter by reducing their LA.

The overall mean daily lying time of 70 minutes per day was rather low compared with other studies that reported total time spent recumbent per day at 100 - 150 minutes (Littlejohn and Munro, 1972; Duncan, 1980). An explanation for the low lying times found in our study could be low soil temperatures and high soil humidity, as horses avoid lying down on wet underground, whereas in the summer, lying at high temperatures may have been avoided because this posture appears to be less effective for thermoregulation compared with standing. Resting in a lying position occurred mainly at night, especially before dawn. A similar diurnal pattern was described for lying in free-ranging Przewalski horses (Boyd, 1988; Berger et al., 1999), Haflinger horses (Lamoot and Hoffmann, 2004) and sheep and cattle (Arnold, 1984). During the feeding trial, the TG ponies tended to have lower lying times compared with the CG ponies, but this difference was only significant for three single days. We did not measure resting in a standing position, but this behaviour may have increased in winter, as observed by Duncan (1985) in Camargue horses.

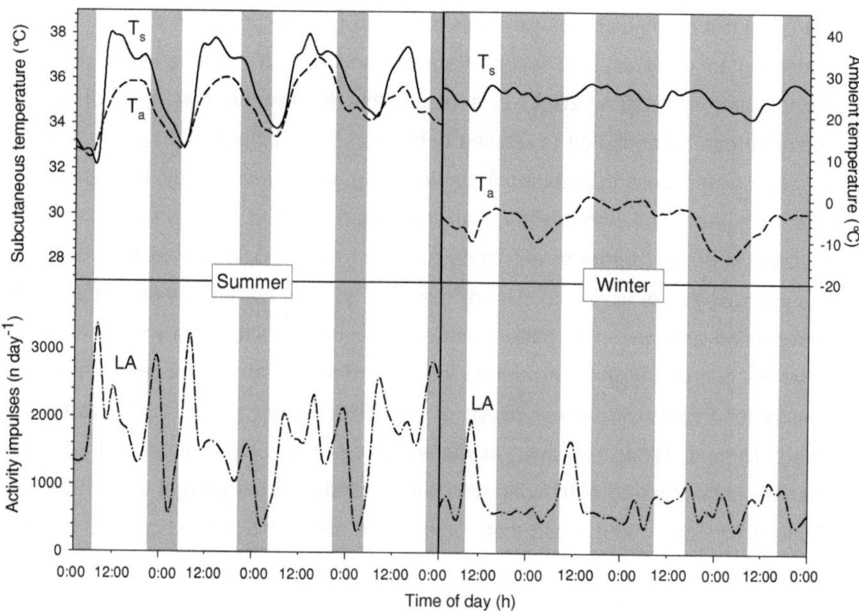

Fig. 4 Examples of the diurnal rhythm of the mean subcutaneous temperature (T_s, solid line), ambient temperature (T_a, dashed line) and locomotor activity (LA, dot-dashed line) in Shetland ponies on four consecutive days in summer (July 2010) and winter (December 2010, across feeding groups). Grey shaded areas indicate scotophase.

Resting in standing position and not in recumbency is the position of minimal energy expenditure (Winchester, 1943); in addition, this posture protects the animal against body cooling through conduction and therefore possibly contributed to energy savings in the TG ponies.

Seasonal changes in physiological parameters

The skin and the subjacent tissues belong to the body shell, and thus T_s usually fluctuates more than T_b (Scheunert and Trautmann, 1987; Schmidt-Nielsen, 1997). With decreasing T_a, the difference between T_s and T_b increases. In goats, the gradient was found to be 1 °C at a T_a of 20 °C and approximately 3 °C at a T_a of 0 °C (Al-Tamimi, 2006; Al-Tamimi, 2007). Körtner and Geiser (Körtner and Geiser, 2000) and Körtner et al. (Körtner et al., 2001) reported a linear relationship between T_b and skin temperature, in sugar gliders and tawny frogmouths, respectively. Skin temperature has also been frequently used as a suitable indicator for T_b in several other species (Taffe, 2011; Brigham, 1992; Audet and Thomas, 1996; Barclay et al., 1996; Körtner and Geiser, 2000). Even though skin temperature is biophysically distinct from T_s, changes in T_s will likely reflect changes in T_b.

The daily mean T_s fluctuation in the present study, with T_s decreasing during the night and rapidly rising after sunset, is consistent with a normal daily shallow hypometabolism (Fig. 4). This phenomenon can be observed during the 24 hourly rhythms of activity and rest in many species (Heldmaier et al., 2004) and leads to a MR reduction of up to 20% compared with activity phases. In our study, these daily T_s variations occurred over the entire year in close relation to the daily photoperiod (Scheunert and Trautmann, 1987), similar to what other studies have found in horses and red deer (Piccione et al., 2002; Arnold, 2004). The diurnal T_b rhythm is an endogenous rhythm that persists even when exogenous factors have been removed (Scheunert and Trautmann, 1987; Piccione et al., 2002). This explains why T_s also decreased at night on days when T_a did not (Fig. 5). Furthermore, the increase of T_s in the morning tended to correlate with an increase in LA, especially in summer (Fig. 4), suggesting that the animals possibly used their LA to increase their T_b in the morning. The lower T_s variations in winter and spring (Fig. 3A, 4) may indicate that our animals shifted from a short daily hypometabolism to a more intense hypometabolism to save energy. This assumption is supported by our results for heart rates (Fig. 2), which decreased significantly during winter and spring compared with summer and thus indicate a reduced energy expenditure. However, even though daily hypometabolism would account for the observed rhythmic diurnal changes in T_s, it does not explain the large daily T_s variations observed in spring and especially in summer, or the seasonal changes of mean daily T_s. A possible explanation is that in summer, horses may have less need to keep their T_b constant during the day because there is plenty of food and T_a values are moderate to high. However, there are some theoretical studies that predict the exact opposite (e.g. Angilletta et al., 2010). Nevertheless, the reduction of T_b at the night may improve the ponies` capacity to store heat during hot days and thus reduce the energetic cost for thermoregulation (Fuller et al., 2005; Arnold et al., 2006). This type of adaptive heterothermy has been reported in other ungulates such as the eland (Taylor and Lyman, 1967), oryx (Taylor, 1969), giraffe (Langman and Maloiy, 1989) and Thomson's as well as Grant's gazelles (Taylor, 1970).

In our study, the variation in T_s abruptly decreased in the winter months (Fig. 3A, Fig. 4). Because T_s can be lowered by peripheral vasoconstriction to protect the body core against heat loss at times of low T_a (Schmidt-Nielsen, 1997, Arnold et al., 2006), the lowest T_s values are expected to coincide with the lowest T_a. However, even though T_s and T_a were found to be correlated in the winter months, the lowest mean T_s and $T_{s,min}$ were recorded in March and April long after the winter T_a nadir in December. Moreover, $T_{s,min}$ was negatively correlated with T_a, indicating that variations of T_s were therefore not induced by variations of T_a but by active thermoregulation. Thus the decrease in the variation of T_s can be explained to some extent by an adaptation to reduce energy. Similar mechanisms were described recently in kangaroos (Maloney et al., 2011).

Measuring T_r is a reliable method to measure T_b of animals (Scheunert and Trautmann, 1987; Green et al., 2005). The normal range of T_b in horses is 37.5 to 38.5 °C (Scheunert and Trautmann, 1987), and a deviation from the normothermia of more than 1 °C is considered to lead to discomfort, and an increase of 5 °C or a decrease of 10 °C will be fatal (Langlois, 1994). The mean T_r of 36.4 °C in the TG ponies measured at the end of the second winter period was more than 1 °C below the lower T_b limit and significantly lower than the T_b of the CG mares. Furthermore, T_s of TG animals was lower compared with that of CG animals on most of the days during the feeding trial, even though T_a values were the same for both groups. These T_s differences between both groups were especially pronounced during cold winter days (i.e. $T_a < 0$ °C), and on some days exceeded 2.5 °C (Fig. 5), indicating that TG animals allowed their T_b to drop further than CG animals, thus saving more energy. Alternatively, but less likely, the difference in Ts could be also due to energy limitations. However, lowering the Tb during adverse environmental conditions must involve some trade-off that is less energy demanding, including the cost for re-warming, than keeping the Tb constant (Angilletta et al., 2010; Boyles et al., 2011). Heart rate is closely related to the oxygen consumption (Butler et al., 2004) and is therefore a reliable indicator of MR (Renecker and Hudson, 1985; Bevan et al., 1995; Woakes et al., 1995; Brosh et al., 1998; McCarron et al., 2001). The high correlation of heart rate with LA can be explained as an automatic increase of heart rate with increasing LA to ensure the oxygen supply of the muscle metabolism. However, mean heart rate started to decrease over 1 month after the LA decline (Figs 2 and 3B) and paralleled reductions in variations in body mass, T_s and $T_{s,min}$. Considering these findings, we suggest that reduced food availability and subsequent loss in body mass initiated energy-saving mechanisms such as a reduction in MR, indicated by a reduced heart rate.

Seasonal changes in metabolic rate

The measured behavioural and physiological parameters show clear seasonal fluctuations, supporting the view of an underlying endogenous seasonal cycle of the MR in Shetland ponies. Similar seasonal patterns of T_s and $T_{s,min}$ were observed by Arnold et al. (Arnold et al., 2004; Arnold et al., 2006) in Przewalski horses and red deer, where heart rate decreased with T_s. Because heart rate is an indicator of MR (Butler et al., 2004), the authors interpreted these reductions as a nocturnal hypometabolism and defined it as an intensified daily hypometabolism occurring when little food is available and the body energy reserves become depleted in late winter. In our study, reduced T_s variations coincided with a rapid drop in heart rate, both possibly indicating a reduction in MR. Reduced heart rate may also be caused by reduced LA, which simultaneously leads to reduced heat production and thus lower T_s. However, the LA had no significant correlation with T_s and the start of the decrease in heart rate did not coincide with that of LA. Therefore, the reduction in T_s variations in winter might reflect the protection of the body against cooling because re-warming by locomotion is highly energy demanding during periods of low T_a and low food supply, and thus would deplete the energy reserves more quickly.

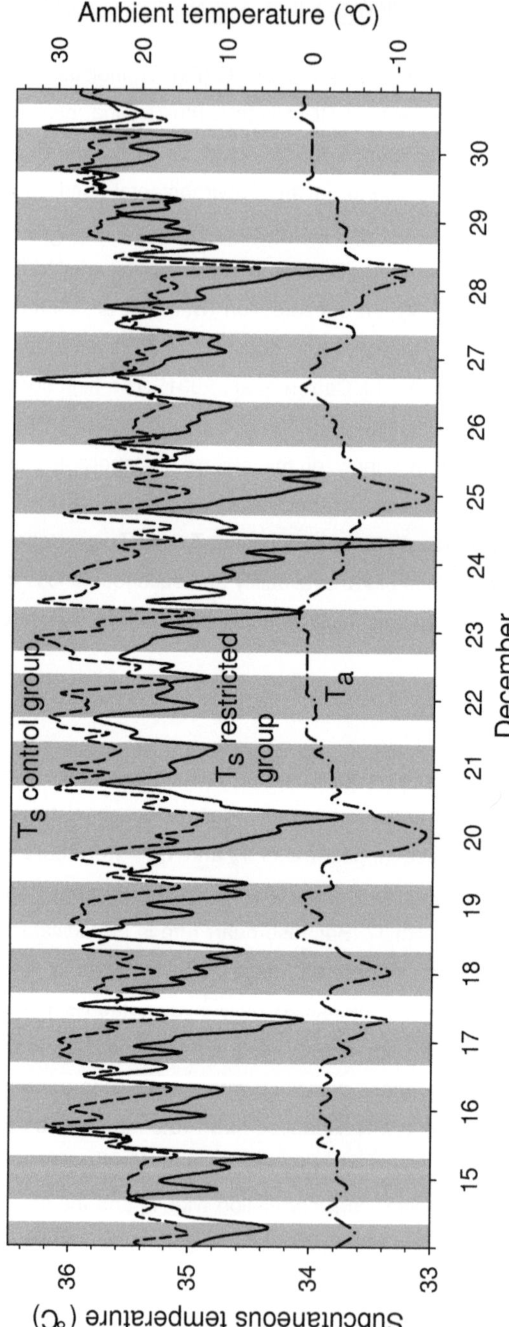

Fig. 5 Example of the diurnal rhythm of ambient temperature (T_a, dot-dashed line) and the mean subcutaneous temperature (T_s) in winter of restrictively fed ($N = 5$, solid line) and control fed ($N = 5$, dashed line) Shetland ponies (see Materials and methods for details). Grey shaded areas indicate scotophase.

The restrictive feeding during the second winter period resulted in a reduced heart rate in TG animals, suggesting a reduction in MR compared with CG animals. The measured heart rate of 26 beats min^{-1} at the end of the trial (February 2011) was below the standard value for horses (30–40 beats min^{-1}). Furthermore, the standard heart rate value in ponies can be expected to be higher than that of in horses, as smaller animals generally have higher heart rate frequencies (Stahl, 1967).

Influence of domestication

In our study, domesticated Shetland ponies showed seasonal adjustment mechanisms similar to those described for their wild counterparts, the Przewalski horse (Arnold et al., 2006; Kuntz et al., 2006), including diurnal hypometabolism and, presumably, winter hypometabolism. These results indicate that this horse breed has maintained the capacity for seasonal adaptation mechanisms during domestication. However, the feeding trial in the second winter also revealed a striking difference with available reports on northern wild herbivores (Kuntz et al., 2006). Contrary to the observation in Przewalski horses (Arnold et al., 2006), the CG ponies in the present study did not reduce their feed intake. Instead, animals concomitantly increased their body mass, T_r and heart rate. A possible explanation is that the heat increment of feeding masked an underlying seasonal cycle of MR (Worden and Pekins, 1995). The results of our study clearly contradict speculations that some livestock species have lost their ability to reduce their energy expenditure under circumstances of food shortage (Price, 1984), as they were housed and selected under constant nutrient supply without selection pressure for maintaining a seasonal, cyclic adaptation. In contrast, the seasonal fluctuations of the measured parameters are much in line with previous findings in wild horses or ungulates (Arnold et al., 2004; Arnold et al., 2006). However, Shetland ponies show metabolic features that indicate higher metabolic efficiency (thrifty genotypes) (Kienzle et al., 2010) that can lead to adiposity (Geor, 2008) and physical inactivity (Prentice et al., 2005) under high feeding intensity. In addition, they develop a very thick hair coat in winter and therefore resemble wild horses more than purebred horses selected for high locomotor efficiency. Shetland ponies can be characterized as a primary population after the first steps of domestication according to the domestication model of Lauvergne (Lauvergne, 1994). We suggest that during domestication, several stepwise changes might have occurred. During quite an early stage of domestication, human selection pressure might have acted against seasonal cycles of endogenous control in appetite, thus allowing domestic animals to develop body reserves even during winter times. However, at least in a robust breed like the Shetland pony, the capability to reduce the MR during winter was not lost during domestication. We therefore conclude that domestication had little effect on the Shetland ponies' adjustment mechanisms in relation to environmental conditions, as our data show clear signs of a hypometabolism.

Acknowledgement

The authors thank Jürgen Dörl for technical help and for taking care of the animals.

References

Al-Tamimi, H. J. (2006). Responses of core and peripheral temperatures to chronic cold stress in transiently goitrous goats. *J. Therm. Biol.* 31, 626-633.

Al-Tamimi, H. J. (2007). Responses of simultaneously recorded intraperitoneal and subcutaneous temperatures of Black Bedouin goats to transient thyrosuppression during cold stress. *Livest. Sci.* 106, 254-260.

Angilletta, M.J., Cooper, B.S., Schulter, M.S. and Boyles, J.G. (2010). The evolution of thermal physiology in endotherms. *Front. Biosci.* E2, 861-881.

Arnold, G. W. (1984). Comparison of the time budgets and circadian patterns of maintenance activities in sheep, cattle and horses grouped together. *Appl. Anim. Behav. Sci.* 13, 19-30.

Arnold, W., T. Ruf and R. Kuntz (2006). Seasonal adjustment of energy budget in a large wild mammal, the Przewalski horse (*Equus ferus przewalskii*) II. Energy expenditure. *J. Exp. Biol.* 209, 4566-4573.

Arnold, W., T. Ruf, S. Reimoser, F. Tataruch, K. Onderscheka and Schober, F. (2004). Nocturnal hypometabolism as an overwintering strategy of red deer (*Cervus elaphus*). *Am. J. Physiol. Reg I* 286, R174-R181.

Aschoff, J. (1962). Spontane lokomotorische Aktivität. De Gruyter Verlag, Berlin, Germany.

Audet, D. and D. W. Thomas (1996). Evaluation of the accuracy of body temperature measurement using external radio transmitters. *Can. J. Zool.* 74, 1778-1781.

Barclay, R. M. R., M. C. Kalcounis, L. H. Crampton, C. Stefan, M. J. Vonhof, L. Wilkinson and R. M. Brigham (1996). Can external radiotransmitters be used to assess body temperature and torpor in bats? *J. Mammal.* 77, 1102-1106.

Berger, A., K. M. Scheibe, A. Brelurut, E. Schober and W.Streich (2002). Seasonal variation of diurnal and ultradian rhythms in red deer. *Biol. Rhythm Res.* 33, 237-253.

Berger, A., K. M. Scheibe, K. Eichhorn, A. Scheibe and J. Streich (1999). Diurnal and ultradian rhythms of behaviour in a mare group of Przewalski horse (*Equus ferus przewalskii*), measured through one year under semi-reserve conditions. *Appl. Anim. Behav. Sci.* 64, 1-17.

Berger, A., K. M. Scheibe, K. Wollenweber, B. Patan, P. Schnitker, C. Herrman and K. D. Budras (2006). Jahresrhythmik von Aktivität, Nahrungsaufnahme,

Lebendmasse und Hufentwicklung bei Wild- und Hauspferden in naturnahen Lebensbedingungen. In: *Aktuelle Arbeiten zur artgemäßen Tierhaltung 2006* (ed. Einschütz, K.), pp 137-146, KTBL Schrift 448, KTBL, Darmstadt.

Bevan, R. M., J. R. Speakman and P. J. Butler (1995). Daily energy-expenditure of tufted ducks - a comparison between indirect calorimetry, doubly labeled water and heart-rate. *Funct. Ecol.* 9, 40-47.

Boyd, L. E. and N. Bandi (2002). Reintroduction of takhi, *Equus ferus przewalskii*, to Hustai National Park, Mongolia, time budget and synchrony of activity pre- and post-release. *Appl. Anim. Behav. Sci.* 78, 87-102.

Boyd, L. E. (1988). Time budgets of adult Przewalski horses - effects of sex, reproductive status and enclosure. *Appl. Anim. Behav. Sci.* 21, 19-39.

Boyd, L. E., D. A. Carbonaro and K. A. Houpt (1988). The 24-hour time budget of Przewalski horses. *Appl. Anim. Behav. Sci.* 21, 5-17.

Boyles J. G., Seebacher, F., Smit, B. and McKechnie, A. E. (2011). Adaptive thermoregulation in endotherms may alter response to climate change. *Integr. Comp. Biol.* 51, 676-690.

Brigham, R. M. (1992). Daily torpor in a free-ranging goatsucker, the common poorwill (*Phalaenoptilus nuttallii*). *Physiol. Zool.* 65, 457-472.

Brosh, A., Y.Aharoni, A. A. Degen, D. Wright and B. Young (1998). Estimation of energy expenditure from heart rate measurements in cattle maintained under different conditions. *J. Anim. Sci.* 76, 3054-3064.

Butler, P. J., J. A. Green, I. L. Boyd and J. R. Speakman (2004). Measuring metabolic rate in the field, the pros and cons of the doubly labelled water and heart rate methods. *Funct. Ecol.* 18, 168-183.

Carroll, C. L. and P. J. Huntington (1988). Body condition scoring and weight estimation of horses. *Equine Vet. J.* 20, 41-45.

Duncan, P. (1980). Time budgets of Camargue horses. 2. Time-budgets of adult horses and weaned sub-adults. *Behaviour* 72, 26-49.

Duncan, P. (1985). Time-budgets of Camargue horses. 3. Environmental-influences. *Behaviour* 92, 188-208.

Fuller, A., P. R. Kamerman, S. K. Maloney, A. Matthee, G. Mitchell and D. Mitchell (2005). A year in the thermal life of a free-ranging herd of springbok *Antidorcas marsupialis*. *J. Exp. Biol.* 208, 2855-2864.

Geiser, F. (1988). Reduction of metabolism during hibernation and daily torpor in mammals and birds - temperature effect or physiological inhibition. *J. Comp. Physiol. B* 158, 25-37.

Geiser, F. and T. Ruf (1995). Hibernation versus daily torpor in mammals and birds - physiological variables and classification of torpor patterns. *Physiol. Zool.* 68, 935-966.

Geor, R. J. (2008). Metabolic predispositions to laminitis in horses and ponies, obesity, insulin resistance and metabolic syndromes. *J. Equine Vet. Sci.* 28, 753-759.

Gesellschaft für Ernährungsphysiologie (1994). Empfehlung zur Energie - und Nährstoffversorgung der Pferde. DLG-Verlag, Frankfurt a.M., Germany.

Green, A. R., S. G. Gates and L. M. Lawrence (2005). Measurement of horse core body temperature. *J. Therm. Biol.* 30, 370-377.

Heldmaier, G., S. Ortmann and R. Elvert (2004). Natural hypometabolism during hibernation and daily torpor in mammals. *Resp. Physiol. Neurobiol.* 141, 317-329.

Heldmaier, G., S. Steinlechner, T. Ruf, H. Wiesinger and M. Klingenspor (1989). Photoperiod and thermoregulation in vertebrates body temperature rhythms and thermogenic acclimation. *J. Biol. Rhythm* 4, 251-265.

Kienzle, E., M. Coenen and A. Zeyner (2010). Der Erhaltungsbedarf von Pferden an umsetzbarer Energie. *Übersichten Tierernährung* 38, 33-54.

Körtner, G., R. M. Brigham and F. Geiser (2001). Torpor in free-ranging tawny frogmouths (*Podargus strigoides*). *Physiol. Biochem. Zool.* 74, 789-797.

Körtner, G. and F. Geiser (2000). Torpor and activity patterns in free-ranging sugar gliders *Petaurus breviceps (Marsupialia)*. *Oecologia* 123, 350-357.

Kuntz, R., C. Kubalek, T. Ruf, F. Tataruch and W. Arnold (2006). Seasonal adjustment of energy budget in a large wild mammal, the Przewalski horse (*Equus ferus przewalskii*) I. Energy intake. *J. Exp. Biol.* 209, 4557-4565.

Lamoot, I. and M. Hoffmann (2004). Do season and habitat influence the behaviour of Haffinger mares in a coastal dune area? *Belg. J. Zool.* 134, 97-103.

Langlois, B. (1994). Inter-breed variation in the horse with regard to cold adaptation - a review. *Livest. Prod. Sci.* 40, 1-7.

Langman, V. A. and G. M. O. Maloiy (1989). Passive obligatory heterothermy of the giraffe. *J. Physiol. (London)* 415, P89-P89.

Lauvergne, J. J. (1994). Characterization of domesticated genetic resources in American camelids, a new approach, In: *Proceedings of the European Symposium of South American Camelides* (eds. M Gerken, C. Renieri), pp 59-65.

Littlejohn, A. and R. Munro (1972). Equine recumbency. *Vet. Rec.* 90, 83-85.

Malechek, J. C. and B. M. Smith (1976). Behavior of range cows in response to winter weather. *J. Range Manag.* 29, 9-12.

Maloney, S. K., A. Fuller, L. C. R. Meyer, P. R. Kamerman, G. Mitchell and D. Mitchell (2011). Minimum daily core body temperature in western grey kangaroos decreases as summer advances, a seasonal pattern, or a direct response to water, heat or energy supply? *J. Exp. Biol.* 214, 1813-1820.

McCarron, H. C. K., R. Buffenstein, F. D. Fanning and T. J. Dawson (2001). Free-ranging heart rate, body temperature and energy metabolism in eastern grey kangaroos (*Macropus giganteus*) and red kangaroos (*Macropus rufus*) in the arid regions of South East Australia. *J. Comp. Physiol. B* 171, 401-411.

McFarland, D. (1985). Animal Behaviour. London: Addison Wesley Longman Limited.

Piccione, G., G. Caola and R. Refinetti (2002). The circadian rhythm of body temperature of the horse. *Biol. Rhythm Res.* 33, 113-119.

Prentice, A. M., P. Rayco-Solon, and S. E. Moore (2005). Insights from the developing world, thrifty genotypes and thrifty phenotypes. *Proc. Nutr. Soc.* 64, 153-161.

Price, E. O. (1984) Behavioral aspects of animal domestication. *Quart. Rev. Biol.* 59, 1-32.

Renecker, L. A. and R. J. Hudson (1985). Telemetered heart-rate as an index of energy-expenditure in moose (*Alces alces*). *Comp. Biochem. Physiol. A* 82, 161-165.

Statistical Analysis System. (2008). User's guide 9.2. Cary: NC, SAS Institute Inc.

Scheibe, K. M., A. Berger, J. Langbein, W. J. Streich and K. Eichhorn (1999). Comparative analysis of ultradian and circadian behavioural rhythms for diagnosis of biorhythmic state of animals. *Biol. Rhythm. Res.* 30, 216-233.

Scheunert, A. and A. Trautmann (1987). Lehrbuch der Veterinär-Physiologie. Verlag Paul Parey, Berlin Hamburg, Germany.

Schmidt-Nielsen, K. (1997). Animal Physiology - Adaptation and Environment. Cambridge University Press, Cambridge, UK.

Singer, D. (2007). Why 37 degrees °C? Evolutionary fundamentals of thermoregulation. *Anaesthesist* 56, 899-906.

Stahl, W. R. (1967). Scaling of respiratory variables in mammals. *J. Appl. Physiol.* 22, 453-460.

Taffe, M. A. (2011). A comparison of intraperitoneal and subcutaneous temperature in freely moving rhesus macaques. *Physiol. Behav.* 103, 440-444.

Taylor, C. R. (1969). The eland and the oryx. *Sci. Am.* 220, 88-95.

Taylor, C. R. (1970). Strategies of temperature regulation - effect on evaporation in East African ungulates. *Am. J. Physiol.* 219, 1131 - 1135.

Taylor, C. R. and C. P. Lyman (1967). A comparative study of environmental physiology of an East African antelope, the eland, and the Hereford steer. *Physiol. Zool.* 40, 280-295.

Weiner, J. (1977). Energy-metabolism of roe deer. *Acta Theriol.* 22, 3-24.

Winchester, C. F. (1943). The energy cost of standing in horses. *Science* 97, 24.

Woakes, A. J., P. J. Butler and R. M. Bevan (1995). Implantable data logging system for heart-rate and body-temperature - its application to the estimation of field metabolic rates in Antarctic predators. *Med. Biol. Eng. Comp.* 33, 145-151.

Worden, K. A. and P. J. Pekins (1995). Seasonal change in feed intake, body composition, and metabolic rate of white-tailed deer. *Can. J. Zool.* 73, 452-457.

CHAPTER 3

Effect of long-term feed restriction on health status and welfare of a robust horse breed, the Shetland pony mares (*Equus ferus caballus*)

L. Brinkmann, M. Gerken

Department of Animals Sciences, University of Goettingen, Albrecht-Thaer Weg 3, 37075 Goettingen, Germany

Effect of long-term feed restriction on health status and welfare of a robust horse breed, the Shetland pony mares (*Equus ferus caballus*)

Lea Brinkmann, Martina Gerken, Alexander Riek

Department of Animal Sciences, University of Goettingen, Albrecht-Thaer-Weg 3, 37075 Goettingen, Germany

Abstract

Outdoor group housing is increasingly recognized as an appropriate housing system for domesticated horses. The objective of this study was therefore to investigate the effect of potential feed shortage in semi-extensive horse keeping systems simulated by feed restriction in winter on animal health and welfare. In ten female non pregnant Shetland ponies blood concentrations of NEFA, total protein (TP), total bilirubin (TB), beta-hydroxybutyrate (BHB) and thyroxine were monitored on a monthly basis for seven month. Body mass and body condition score (BCS) were also recorded. After three months of free feed availability on rye grass dominated pasture, animals were transferred to open stables and allocated into one control and one feed restriction group of five animals each. After 14 weeks of feed restriction, mean body mass loss was 18.4 ± 2.99% and the BCS decreased by 2.2 ± 0.8 points (BCS scale: 0 = emaciated, 5 = obese). In the control group body mass and BCS remained constant. Feed restriction led to a continuous increase in TB ($P < 0.001$) and NEFA ($P < 0.01$) concentrations compared to control ponies. The TP and BHB values only differed at the end of the trial with lower concentrations in restricted mares ($P < 0.05$). Feed restriction had no effect on thyroxine concentrations. Among the blood parameters measured, TB concentrations in the feed restricted group were out of the reference range during the entire feeding trial, indicateing an excessive demand on the liver capacity. TP concentrations instead were out of the reference range only during the last two weeks of the feeding trial. The increased NEFA concentration in feed restricted compared to control ponies suggest that fat was mobilized. Our results on changing blood parameters in feed restricted mares indicate beginning health problems. The BCS, as well as plasma NEFA and TB concentrations were good indicators for a rapid detection of possible health problems caused by undernourishment in horses when kept under semi-natural conditions. In contrast, blood parameters of the control animals were within the reference ranges, suggesting that a year round outdoor housing with additional feed supply is an adequate housing system for a robust horse breed like the Shetland pony.

Keywords: animal welfare, blood parameter, extensive housing, feed restriction, horse, winter conditions

Introduction

Outdoor group housing is increasingly recognized as an appropriate housing system for domesticated horses (Zeeb and Schnitzer, 1997). However, welfare impairments of the horses when kept under harsh environments are controversially discussed e.g. by Robinson et al. (2006). In the Northern hemisphere ambient temperatures (Ta) can vary substantially over the course of the year, thus exposing animals to extreme environmental conditions. As an adaptation mechanism, behavioural as well as endogenous adjustments sustain homoeostasis (Arnold and Dudzinski, 1978; McFarland, 1985). Wild horses are described to adapt to changing environmental conditions, e.g. by their feeding behaviour (Cosyns et al., 2001), the accumulation of body fat for cold winter times when little feed is available (Scheibe and Streich, 2003) or even by reducing body temperature, heart rate and metabolic rate (Arnold *et al.*, 2006; Brinkmann et al., 2012).

Several studies investigated the effects of food restriction in horses (Dugdale et al., 2010; Powell et al., 2000; McManus and Fitzgerald, 2000). In particular, blood parameters have been shown to provide essential information on the physiological state of horses (Sticker *et al.*, 1995a, b; Christensen *et al.*, 1997). However, in these studies horses were starved or feed restricted for a relatively short time. To our knowledge, there are no reports on the effects of a long term controlled feed restriction in horses kept outdoors under varying Ta's comparable to those experienced by wild equids under natural conditions. We simulated winter feed shortage found in natural habitats by gradually reducing the energy and protein requirements for some of the ponies to 70%, while for another group of ponies feed was available *ad libitum*. The aim of this study was to investigate the effect of long-term thermal and nutritional stress as well as possible impacts on health of a robust horse breed. Different blood parameters were recorded to evaluate their suitability for evaluation of emerging metabolic disorders and to determine suitable for evaluation of emerging metabolic disorders and to determine suitable indicators for the early detection of possible welfare and health problems in horses kept under these harsh conditions.

Materials and Methods

Procedures performed in this study were in accordance with the German animal ethics regulations and approved by the State Office of Lower Saxony for Consumer Protection and Food Savety, Germany (Ref. no.: 33.14-42502-04-083/09).

Animals and study site

The study was undertaken at the Department of Animal Science at the University of Goettingen, Germany, from August 2010 until February 2011.

The experiment involved 10 female non pregnant Shetland ponies (age: 4 to 16 years, initial body mass: 110 to 228 kg; initial BCS: 3.0 – 4.5 points, BCS scale: 0 = emaciated, 5 = obese). All animals originated from different farms in Germany and the Netherlands and were accustomed to outdoor housing systems.

From the 1st of August until the 13th of October 2010 the animals were kept together on a permanent pasture with access to 2 tents (4.0 x 3.6 m) with straw bedding. At the 13th of October animals were taken to the research stable at the Department of Animal Sciences of the University of Goettingen, where they were allocated into a control and a treatment group of 5 animals each, with comparable mean BCS (restricted group: 4.5 ± 0.4 points, control group: 3.8 ± 0.6 points), body mass (restricted group: 162 ± 41 kg; control group: 154 ± 39 kg) and age (restricted group: 8.8 ± 4.9 years; control group: 7.2 ± 2.9 years) for both groups. Despite efforts to match the groups, the inclusion of one very obese animal (body mass: 229 kg; BCS: 4.5) in the restricted group led to a slightly, but not significantly ($P > 0.05$) higher mean body mass and BCS in the feed restricted group compared to the control group. The 2 groups were housed separately in 2 identical rectangular paddocks (210 m2) without vegetation. The paddock was accessible at all times and the surface was covered with wood chips. Each paddock had a permanent access to a pen (6.40 x 2.95 m) with two large exits allowing the animals to enter and leave without rank conflicts. Furthermore both pens were equipped with 5 separate feeding stands (1.35 m x 1.45 m x 0.55 m, height x length x width), one for each pony. The pen floor for all animals was covered by straw until the feeding trial started, when straw was partly replaced by wood ships in the feed restricted group. Pens were not heated and thus temperatures inside the pen were comparable to outside Ta. The dark-light cycle inside the pen fluctuated according to the natural photoperiod. Both groups were given an acclimation period of two weeks before the start of the feeding trial.

Feeding

Animals had free access to water throughout the experiment. During the summer on pasture (August – mid October) straw from the two shelter beddings and a salt lick were available ad libitum (Eggersmann Mineral Leckstein, Heinrich Eggersmann GmbH & Co KG, Rinteln, Germany). Furthermore, in summer all animals had access to a pasture with rye grass dominated grassland. During September and October, ponies were offered additionally approximately 0.5 kg of hay per day and animal. During the two weeks acclimation period before the start of the feeding trial (13 October - 1 November) both groups were fed identically with straw ad libitum, 2 kg hay /100 kg body mass/day and 580 g concentrate/100 kg body mass/day (Derby® Standard, Derby Spezialfutter GmbH, Münster, Germany). In addition, to cover the demand for minerals and vitamins ponies received 22 g /100 kg body mass/day of a commercial mineral supply (Derby® Mineralpellets, Derby Spezialfutter GmbH, Münster, Germany). After the acclimation period the feeding trial started lasting 4 months (November – February) with the 2 groups receiving different diets, fed twice daily at 0800 and 1600 h. While the control group was fed as before, the treatment group was fed restrictively to simulate limited feed availability under natural conditions during autumn and winter. The level of feed restriction increased with time to simulate diminishing feed availability in winter under natural conditions. Therefore, the amount of feed offered to the treatment animals was gradually reduced from 100 % to 80 % of the recommended energy and protein requirements for Shetland ponies (GfE, 1994) until 21 Jan and to a further 70 % until 28 Feb. The daily maintenance requirement of energy and protein for small ponies was assumed to be 0.40 MJ ME /kg body mass0.75 /d (Kienzle et al., 2010) and 3 g crude protein (CP) /kg body mass0.75/ d (GfE, 1994), respectively. However, we expected the energetic needs to be 20% higher for ponies kept in groups and under cold climatic conditions (Kienzle et al., 2010) resulting in a maintenance requirement of 0.48 MJ ME/kg body mass0.75/d for our ponies in winter times. Therefore, the restrictively fed animals received 0.6 kg hay/100 kg body mass/d and 2.0 kg wheat straw/100 kg body mass/d (total ~12.5 MJ and 72 g CP/100 kg body mass/d) from the 1 November until 21 January and subsequently 0.4 kg hay/100 kg body mass/d and 2.0 kg straw/100 kg body mass/d (total ~11.0 MJ and 60 g CP/100 kg body mass/d) until the end of the feeding trial.

In both groups feeding took place in the above described feeding stands to ensure that the amount of feed allocated to each animal was consumed. In addition all animals had access to a standard mineral supplement (Derby® Mineralpellets, Derby Spezialfutter GmbH, Münster, Germany) and a frost proof watering place during the entire feeding trial.

Measurements

Body mass was recorded in bi-weekly intervals with a mobile scale (Weighing System MP 800, resolution: 0.1 kg , Patura KG, Laudenbach, Germany) in order to control the effect of the feeding regime and to avoid that ponies lost more than 20% of their initial body mass measured at the beginning of the feeding trial. The BCS, a palpable and visual assessment of the degree of fatness in the neck, back, ribs and pelvis, was taken monthly according to Carroll and Huntington (1988) throughout the study. The T_a and relative humidity were recorded continuously throughout the study with miniature data loggers in the paddock and on the pasture at 10 minutes intervals (Tiny view TV 1500, temperature resolution: 0.25 °C, humidity resolution: 0.5%, Gemini, Chichester, West Sussex, UK).

During the entire experiment, blood samples (5 ml) were taken on monthly basis at the middle of each month from the vena jugularis into blood tubes containing sodium citrate between 0800 and 0900 h on the sampling day. Serum samples were stored at room temperature until the coagulum was formed. All samples were then centrifuged for 10 minutes at 3000 rpm and 20 °C (centrifugation force: 1620 g). The plasma and serum samples were pipetted into 0.7-ml glass vials and stored at –20 °C until analysis. NEFA, beta-hydroxybutyrate (BHB), total protein (TP) and total bilirubin (TB) were analyzed by colometric enzymatic reactions. Total protein and TB were measured using Bilirubin total (NBD) KonelabTM /T Series and total protein plus KonelabTM /T Series. NEFA concentrations were determined using the NEFA-kit of LT-SYS® (Labor und Technik Eberhard Lehmann , Berlin, Germany) and BHB by the Ranbut-kit (3-D-Hydroxybutyrate) produced by Randox Laboratories (Crumlin, County Antrim, UK).

The thyroxine (T4) concentrations were analyzed with a SNAP® Total T4-Test in a SNAPShot Dx® (IDEXX Laboratories, Ludwigsburg, Germany).

Statistical Analysis

All statistical analyses were performed using the software package SAS version 9.2 (SAS Inst. Inc., Cary, NC). For every feeding group monthly averages of NEFA, TB, TP, T_4 and BHB concentrations were calculated. Differences in blood parameters between the feeding groups were tested with a post hoc test (Tukey) within a mixed model (PROC MIXED) to account for repeated measurements of the same individuals. The model was:

$$Y_{ijklm} = \mu + G_i + M_j + (G \times M)_{ij} + A_k + e_{ijk,}$$

where Y_{ijkl} is the corresponding blood parameter; μ is the mean effect; G is the fixed effect of the treatment group (i = control, restricted); M is the fixed effect of the month (j = August to February); G x M is the interaction between the treatment group and the month; A is the random effect of the animal and e_{ijk} is the random error.

Differences between the 2 feeding groups in BW and BCS were calculated using the Mann-Whitney-U test. All values are presented as means ± SD. In addition, linear regressions of parameters (NEFA, TB, TP, T_4 and BHB) recorded over time were computed by feeding groups.

Results

Climate

Ambient temperature varied considerably during the entire experimental period (7 months) and ranged from −18 °C to 30 °C. During the 4 month feeding trial, monthly mean T_a was 6 ± 5 °C (November, range: -4 ± 2 °C – 15 ± 2 °C), -3 ± 3 °C (December, range: -9 ± 3 °C – 5 ± 1 °C), 3 ± 4 °C (January, range: -4 ± 2 °C – 11 ± 1 °C), and 3 ± 4 °C (February, range: -6 ± 3 °C – 10 ± 5 °C), respectively. This experimental period included 44 days with snow with heights over 1 cm and 84 days with ground frost.

Body mass and BCS

From August until October, i.e. before the feeding trial, there were no differences (P > 0.05) between the prospective feeding groups with regard to BW and BCS (Figure 1). With the beginning of feeding ponies restrictively, body mass decreased continuously with an average of mass loss of 0.25 ± 0.07 kg/d (body mass, kg = 165.7 - 4.18 month, R2 = 0.12, P < 0.05), resulting in an average body mass loss of 18.4 ± 2.99% (range: 15 – 23 %) at the end of the experiment (Fig. 1). Body mass loss from the initial mass was 2.7 ± 2.0% in November, 5.2 ± 1.8% in December, 12.6 ± 1.9% in January and 18.2 ± 2.3% in February. Body mass in the control animals instead did not change during the experimental period (body mass, kg = 149.8 - 0.35 month, R2 = 0.05, P = 0.93). The body mass of ponies differed significantly (P < 0.01) between the two feeding groups in January and February (Fig. 1). Similarly, the BCS decreased in feed restricted animals (BCS, points = 5.4 - 0.73 month, R2 = 0.65, P < 0.001) while it remained stable in the control fed animals (BCS, points = 3.4 - 0.09 month, R2 = 0.03, P = 0.43) differing significantly (P < 0.01) between the two groups in January and February (Fig.1).

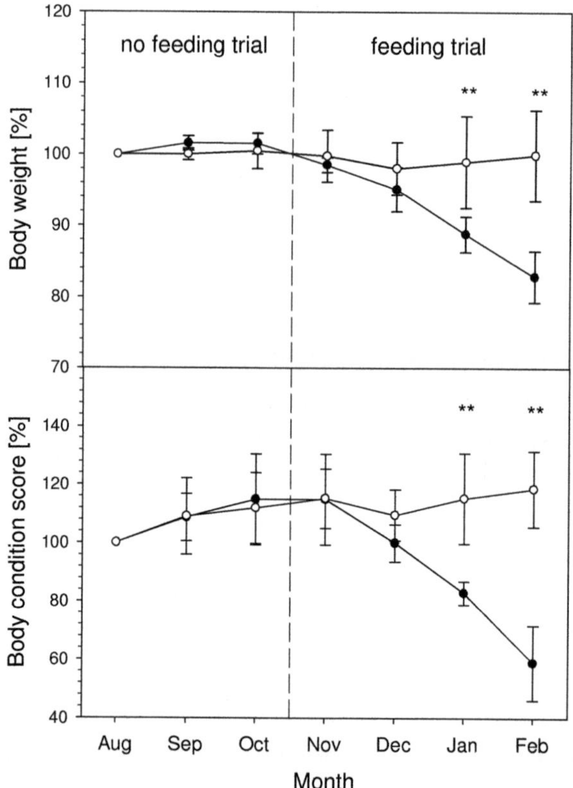

Fig 1. Average monthly body mass (in % of change to initial body mass) and body condition scores (in % of change to initial score) in Shetland pony mares. All animals (circles and dots, n = 10) had access to a pasture and straw from beddings from Aug – Oct ('no feeding trial') and were then fed either restrictively (dots, n = 5) or ad libitum (circles, n = 5) with hay and straw from Nov – Feb ('feeding trial', see text for details; **P < 0.01, denotes differences between feeding groups).

Blood parameters

From August until October when animals had access to a pasture the two prospective feeding groups did not differ in their blood parameters (P > 0.05, Fig. 2 and 3). However, concentrations of TP, BHB and T4 tended to decrease from summer to winter (Fig. 2), whereas NEFA and TB concentrations increased significantly. During the feeding trial TP and BHB concentrations in February differed significantly between both groups after prolonged feed restriction (P < 0.05). T4 concentrations did not differ between the two feeding groups. However, in the feed restricted ponies T4 tended to decrease throughout the feeding trial from 2.5 µg/dl in November to 1.5 µg/dl in February (T4, µg/dl = 3.58 - 0.31 month; R2 = 0.17; P = 0.07), whereas the T4 concentrations in the control group remained rather stable (T4, µg/dl = 1.76 + 0.07 month; R2 = 0.02; P = 0.53) (Fig. 2). In contrast feeding treatment exerted a strong effect on NEFA and serum TB concentrations (Fig. 3). Feed restricted ponies had significantly higher NEFA concentrations throughout the feeding trial compared to the control ponies with a continuous significant increase (NEFA, mmol/l = 0.09 – 0.06 month, R2 = 0.47, P < 0.001). NEFA concentrations of control ponies increased significantly from August until November but then decreased from November to February (NEFA, mmol/l = 0.165 - 0.037 month; R2 = 0.33; P < 0.01). Similar effects of feed restriction were found for serum TB concentrations with increasing values for restrictively fed animals (TB, mg/dl = 0.11 + 0.24 month; R2 = 0.65; P < 0.001) and rather constant values for the control animals (TB, mg/dl = 0.47 + 0.02 month, R2 = 0.04, P > 0.05) (Fig. 3).

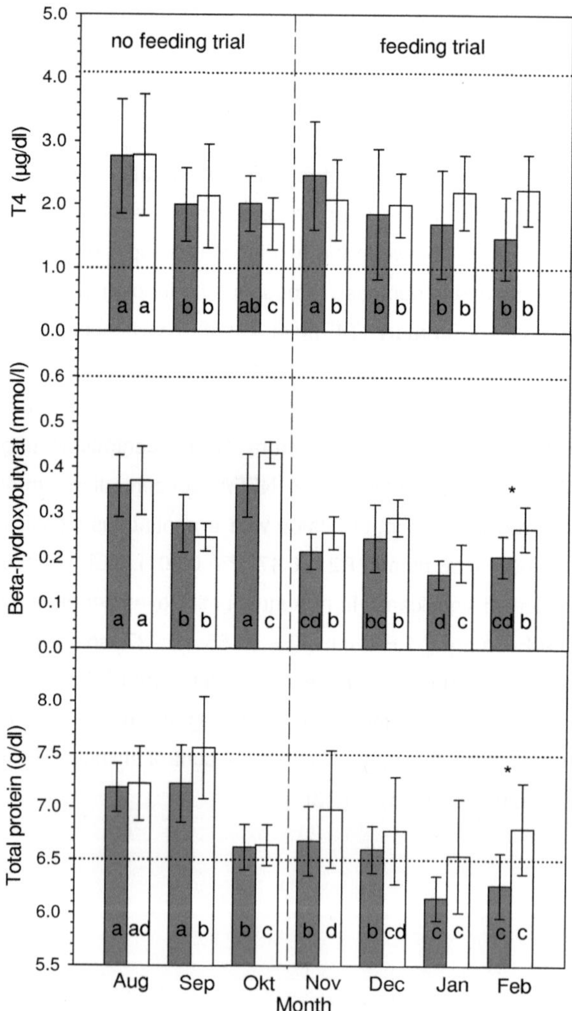

Fig 2. Average monthly plasma concentrations of T4 (µg/dl), beta-hydroxybutyrate (mmol/l) and total protein in Shetland pony mares. All animals (n = 10) had access to a pasture and straw from beddings from Aug – Oct and from then on were either fed restrictively (grey columns, n = 5) or ad libitum (white comuns, n = 5) with hay and straw (Nov – Feb, see text for details; *P < 0.05, ***P < 0.01 denotes differences between feeding groups and columns with different letters differ between months within the same feeding group).

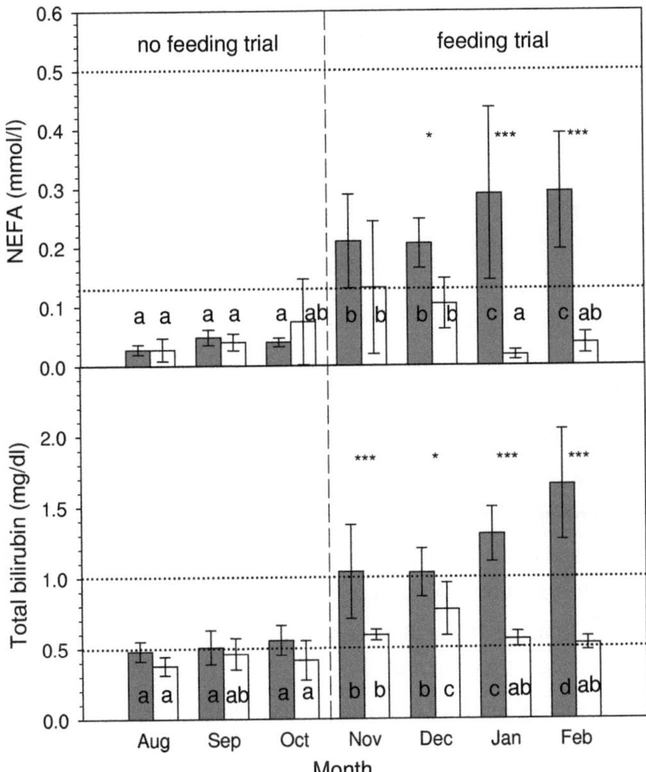

Fig. 3. Average monthly plasma concentrations of non-esterified fatty acids (NEFA, mmol/l) and total bilirubin (mg/dl) in Shetland pony mares. All animals (n = 10) had access to a pasture and straw from beddings from Aug – Oct and from then on were either fed restrictively (grey columns, n = 5) or ad libitum (white comuns, n = 5) with hay and straw (Nov – Feb, see text for details; *P < 0.05, ***P < 0.01 denotes differences between feeding groups and columns with different letters differ between months within the same feeding group).

Discussion

To our knowledge our study is the first analyzing the effects of a long term feed restriction on blood parameters in a robust horse breed under outdoor winter conditions. Our results on Shetland ponies show that feed restriction leads to substantial changes in some blood parameter concentrations directly related to the health status of an animal. In comparison adequately fed animals sustain the same harsh conditions with only little changes in their health status.

Body mass and BCS

While the control group received hay, mineral supply and concentrate in the feeding stands and straw from the bedding, the entire feed of the restricted group was exclusively administered within the feeding stands. All feed restricted ponies lost about 20% of their initial body mass by the end of week 14 of the trial. However, body mass loss slightly preceded any decrease in BCS. Later, both body mass and BCS continued to decrease concomitantly and restricted fed ponies were characterized as fair to thin at the end of the feeding trial according to the BCS recording system from Carroll and Huntington (1988). A slight lag between body mass and BCS change was also observed by Dugdale at al. (2010) and indicates that BSC can be a late indicator of weight loss. This highlights the importance of scoring body condition more frequently if body mass cannot be measured in order to prevent animals becoming too thin too quickly. Additional body fat measurements e.g., belly girths or cresty neck scoring to assess the fat stores within the neck crest might offer additional information of adiposity in the pony. However, data from Dugdale et al. (2010) indicate that crest scoring might be not sufficiently sensitive to detect early weight losses.

Blood parameters

The increase in serum NEFA concentrations found in this study after feed deprivation was consistent with results of other studies in fasted light horses (DePew et al., 1994; Nadal et al., 1997), Welsh ponies (Dugdale et al., 2010), cattle (Rule et al., 1985; Ward et al., 1992) and ewes (Heitmann et al., 1986). It is noteworthy that even after four months of feed restriction under harsh winter conditions, NEFA concentrations in restrictively fed ponies were still in the reference range reported for ponies (0.1 to 0.5

mmol/l; Watson et al., 1992). The central nervous system depends on the energy supply via glucose and physiological mechanisms have been developed to keep blood glucose concentration constant. Inadequate feed intake is usually accompanied by insufficient glucose availability resulting in an increased lipolysis and proteolysis in order to meet energetic needs. Mobilization of lipid stores result in higher plasma NEFA concentrations (Heitmann et al., 1986) as observed in our study. Typically, a small part of NEFA is utilized as an energy source by tissues, however, the bulk is transported to the liver and converted to ketone bodies (e.g. BHB, acetone), which can partly replace glucose in the tissues (Blum et al., 1983). High ketone body concentrations can cause starvation acidosis, but the disposition for ketone body generation in horses is limited so that free fatty acids will primarily be re-esterified into triglycerides. In turn, high triglycerides concentrations can lead to hyperlipidemia (Sjaastad, 2003). Ponies are predisposed for this metabolic disorder frequently accompanied by fatty liver syndrome and accumulation of fat in the kidney, myocardium, and vascular epithelium (Richet and Götze, 1993; Wintzer, 1999; McGavin and Zachery, 2009). Therefore, the lower BHB blood concentrations observed in our study for restrictively fed ponies compared to the control group might be due to the reduced generation of ketone bodies and their high consumption as an alternative to glucose.

Comparing NEFA concentrations during the feeding trial revealed that the NEFA concentrations of the control group at the end of the feeding trial were as low as those of both feeding groups prior to the feeding trial. This indicates that the control animals had no need to mobilize any fat reserves at the end of the feeding trial, unlike the restricted fed animals. The short-term increase of NEFA observed in the control group in November and December may have been caused by the higher energy demand due to low Ta's. The increased NEFA blood concentrations in the restricted group were most likely caused by lipolysis. However, it is known that NEFA is competing with TB for carrier proteins in the blood (e.g. albumin) and transport proteins in the hepatocytes (Kraft and Dürr, 2005) which can lead to hyperbilirubinemia as described previously for ponies and horses (Gronwall and Mia, 1972) and (Naylor et al., 1980). Considering that the normal range of TB concentrations in ponies (up to 1.0 mg/dl) is lower than in other horse breeds (Kraft and Dürr, 2005), we can assume hyperbilirubinemia occurred in our restrictively fed ponies. Thus, our TB values in restrictively fed ponies after 4 months of feed restriction were lower (1.7 mg/dl) compared to TB values reported for feed restricted horses (5.3 mg/dl; Naylor et al., 1980). While Icterus in horses is known to occur at concentrations above 2.0 mg/dl (McGavin and Zachery, 2009), no specifications are known for ponies.The reason for the observed moderate increase in TB concentrations above the critical value in the treatment group might be due to the fact that our ponies were not starved but fed restrictively. Thus, animals did not have to rely on body reserves alone to meet their energetic needs.

In the restrictively fed animals plasma TP concentrations decreased with time and fell even below the lower critical value of 6.5 g/dl (Knickel et al., 2002) at the end of the trial. Total protein consists mainly of albumin and not only serves as a transport mechanism for bilirubin from the blood to the liver but also binds free fatty acids, which if they occur in large quantities in the blood can displace bilirubin from its binding site (Starinsky et al., 1970). Thus, both reduced TP and increased NEFA concentrations can lead to a reduction of bilirubin binding to albumin. These 'free' bilirubin molecules could then penetrate cells and cause cellular damage (Böcker et al., 2008). Therefore, the observed reduction in TP plasma concentrations in our study may be due to malnutrition (Kraft and Dürr, 2005) as well as to a reduced protein production caused by an 'overworking' of the liver due to increased exposure with lipolysis products.

Reduced thyroid hormone concentrations can trigger hypothyroidism (Kraft and Dürr, 2005) resulting in reduced metabolic activity. In our study, long term feed restriction under low Ta's did not influence T4 concentrations. The measured T4 concentrations in our study were at all times and irrespective of the feeding regimen within the reference range for T4 (1.0 - 4.1 µg/dl; Kraft and Dürr, 2005) which is in agreement with results from other horse breeds (Sticker et al., 1995b; Glade and Reimers, 1985). The constant concentration of T4 may be explained by a modified or reduced hepatic conversion of T4 to the metabolically more active T3 (Christensen et al., 1997, van Weyenberg et al., 2008a, Suda et al. 1978).

Conclusions

In our study body mass losses of 1.4% per day in restrictively fed animals over a period of 14 weeks led to distinct changes in some blood parameters, i.e. TP and TB plasma concentrations were outside their reference range (6.5 - 7.5 g/dl and 0 - 1.0

mg/dl, respectively), indicating health problems especially concerning the liver (Kronfeld, 1993). Although, obvious health impairments were not observed in restrictively fed ponies, they became more and more lethargic compared to adequately fed animals, similar to results described for donkeys, mules (Gupta et al., 1999) and sheep (Naqvi and Rai, 1988). This behavioral change of reduced activity could be explained as an energy conservation strategy (Gupta et al., 1999) and may also be interpreted as an indicator for reduced welfare.

The comparison between summer and winter data shows that ponies were able to accumulate body fat reserves in summer which could be mobilized in winter for energy supply. Feed restriction for 14 weeks led to body mass losses of about 20 %, close to the reported threshold of about 25 to 50%, when death from starvation occurs (Engelhardt and Breves, 2000). The rapid mobilization of body fat under feed scarcity, can lead to health problems especially in ponies that tend to show signs of hyperlipidemia as shown by Geor (2008). If horses are kept under extreme outdoor conditions where body mass cannot be monitored regularly, animals should be subjected to a regular BCS to counteract weight loss at an early stage. In this context, crest scoring systems (Znamirowksa, 2005; Carter et al., 2009) provide a suitable superficial index of fat stores in horses. However, scores should then be determined by palpation rather than visual scoring particularly in ponies that develop a thick winter hair coat. Our results on ponies clearly show that among the blood parameters studied, NEFA and TB plasma concentrations are suitable preventive indicators for an early detection of body fat mobilization and resulting possible health problems as already described for cattle (Rule et al., 1985) and sheep (Heitmann et al., 1986).

In the control mares, outdoor housing under harsh winter conditions with adequate feed supply did not lead to changes in blood parameters. Hence, we suggest that in consideration of these parameters, a year round outdoor housing, with regular control of the animals and additional feed supply, can be an adequate housing system for a robust horse breed like the Shetland pony. However, Shetland ponies show metabolic features that indicate higher metabolic efficiency (thrifty genotypes, Kienzle et al., 2010), a phenomenon first described by Neel et al. (1962) for humans, that can lead to adiposity (Geor, 2008). Ponies therefore resemble wild horses more than horses selected for higher locomotor efficiency. Accordingly, different results are expected for warmblooded or even thoroughbred horses because these breeds normally exhibit lower body fat accumulation and less hair coat insulation in winter (Cymbaluk, 1994; Langlois, 1994).

Acknowledgement

The authors thank Jürgen Dörl for help with data collection and care of the animals. The veterinary advice of the veterinarian Michael Beihsner is highly appreciated.

References

Arnold G. W. and M. L. Dudzinski (1978). Ethology of free-ranging domestic animals. Elsevier Scientific Pub. Co., Amsterdam, Netherlands.

Arnold W., T. Ruf and R. Kuntz (2006). Seasonal adjustment of energy budget in a large wild mammal, the Przewalski horse (*Equus ferus przewalskii*) II. Energy expenditure. *J. Exp. Biol.* 209, 4566-4573.

Blum J. W., P. Kunz, H. Leuenberger, K. Gautschi and M. Keller (1983). Thyroid-hormones, blood-plasma metabolites and hematological parameters in relationship to milk-yield in dairy-cows. *Anim. Prod.* 36, 93-104.

Böcker W., H. Denk, P. U. Heitz and H. Moch (2008). Pathologie. 4th ed. Urban & Fischer, München, Germany.

Brinkmann L., M. Gerken, A. Riek (2012). Adaptation strategies to seasonal changes in environmental conditions of a domestic horse breed, the Shetland pony (*Equus ferus caballus*). *J. Exp. Biol.* 215, 1061-1068.

Carroll C.L. and P.J. Huntington (1988). Body condition scoring and weight estimation of horses. *Equine Vet. J.* 20, 41-45.

Carter R. A., R. J. Geor, S. W. Burton, T. A. Cubitt und P. A. Harris (2009). Apparent adiposity assessed by standardised scoring systems and morphometric measurements in horses and ponies. *The Veterinary Journal* 179, 204-210.

Christensen R. A., K. Malinowski, A. M. Massenzio, H. D. Hafs and C. G. Scanes (1997). Acute effects of short-term feed deprivation and refeeding on circulating concentrations of metabolites, insulin-like growth factor I, insulin-like growth factor binding proteins, somatotropin, and thyroid hormones in adult geldings. *J. Anim. Sci.* 75, 1351-1358.

Cosyns E., T. Degezelle, E. Demeulenaere and M. Hoffmann (2001). Feeding ecology of Konik horses and donkeys in Belgian coastal dunes and its implications for nature management. *Belg. J. Zool.* 131 (Suppl. 2), 111-118.

Cymbaluk N.F. (1994). Thermoregulation of horses in cold, winter weather - a review. *Livest. Prod. Sci.* 40, 65-71.

DePew C. L., D. L. Thompson, J. M. Fernandez, L. S. Sticker and D. W. Burleigh (1994). Changes in concentrations of hormones, metabolites, and amino-acids in plasma of adult horses relative to overnight feed deprivation followed by a pellet-hay meal fed at noon. *J. Anim. Sci.* 72, 1530-1539.

Dugdale A. H. A., G. C. Curtis, P. Cripps, P. A. Harris, C. Mc. G. Argo (2010). Effect of dietary restriction on body condition, composition and welfare of overweight and obese pony mares. *Equine Vet. J.* 42, 600-610.

Geor R. J. (2008). Metabolic predispositions to laminitis in horses and ponies: Obesity, insulin resistance and metabolic syndromes. *J. Equine Vet. Sci.* 28, 753-759.

GfE - Gesellschaft für Ernährungsphysiologie (1994). Empfehlung zur Energie- und Nährstoffversorgung der Pferde. DLG-Verlags GmbH, Frankfurt a. M., Germany.

Glade M. J. and T. J. Reimers (1985). Effects of dietary energy supply on serum thyroxine, tri-iodothyronine and insulin concentrations in young horses. *J. Endocrinol.* 104, 93-98.

Gronwall R. and A. S. Mia (1972). Fasting hyperbilirubinemia in horses. *Digest. Dis. Sci.* 17, 473-476.

Gupta A. K., Y. P. Mamta and M. P. Yadav (1999). Effect of feed deprivation on biochemical indices in equids. *Journal of Equine Science* 10, 33-38.

Heitmann R. N., S. C. Sensenig, C. K. Reynolds, J. M. Fernandez and D. J. Dawes (1986). Changes in energy metabolite and regulatory hormone concentrations and net fluxes across splanchnic and peripheral-tissues in fed and progressively fasted ewes. *J. Nutr.* 116, 2516-2524.

Kienzle E., M. Coenen and A. Zeyner (2010). Der Erhaltungsbedarf von Pferden an umsetzbarer Energie. Übersichten Tierernährung 38, 33-54.

Knickel U. R., C. Wilczek, K. Jöst (2002). MemoVet Praxis-Leitfaden Tiermedizin. Schattauer, Stuttgart, Germany

Kraft W. and U. M. Dürr (2005). Klinische Labordiagnostik in der Tiermedizin. 6th ed. Schattauer, Stuttgart, Germany.

Kronfeld D. S. (1993). Starvation and malnutrition of horses - recognition and treatment. *J. Equine Vet. Sci.* 13, 298-304.

Langlois B. (1994). Inter-breed variation in the horse with regard to cold adaptation - a review. *Livest. Prod. Sci.* 40, 1-7.

McFarland D. (1985). Animal Behaviour. 3rd ed. Addison Wesley Longman Limited, Essex, UK.

McGavin M. D. and J. F. Zachery (2009). Pathologie der Haustiere: Allgemeine, spezielle und funktionelle Veterinärpathologie. Urban & Fischer, München, Germany.

McManus C. J. and B. P. Fitzgerald (2000). Effects of a single day of feed restriction on changes in serum leptin, gonadotropins, prolactin, and metabolites in aged and young mares. *Domest. Anim. Endocrin.* 19, 1-13.

Nadal M. R., D. L. Thompson and L. A. Kincaid (1997). Effect of feeding and feed deprivation on plasma concentrations of prolactin, insulin, growth hormone, and metabolites in horses. *J. Anim. Sci.* 75, 736-744.

Naqvi S. M. K. and A. K. Rai (1988). Effect of fasting on some biochemical-constituents of blood in Avivastra sheep. *Indian J. Anim. Sci.* 58, 1079-1081.

Naylor J. M., D. S. Kronfeld and K. Johnson (1980). Fasting hyperbilirubinemia and its relationship to free fatty-acids and triglycerides in the horse. *P. Soc. Exp. Biol. Med.* 165, 86-90.

Neel J. V. (1962). Diabetes mellitus: A "thrifty" genotype rendered detrimental by "progress"? *Am. J. Hum. Genet.* 14, 353-362.

Powell, D. M., L. M. Lawrence, B. P Fitzgerald, K. Danielsen, A. Parker, P. Siciliano and A. Crum (2000). Effect of short-term feed restriction and calorie source on hormonal and metabolic responses in geldings receiving a small meal. *J. Anim. Sci.* 78, 3107-3113.

Richet J. and R. Götze (1993). Tiergeburtshilfe. 3rd ed. Paul Parey Verlag, Hamburg Berlin.

Robinson N. E., W. Karmaus, S. J. Holcombe, E. A. Carr and F.J. Derksen (2006). Airway inflammation in Michigan pleasure horses: prevalence and risk factors. *Equine Vet. J.* 38, 293-299.

Rule D. C., D. C. Beitz, G. Deboer, R. R. Lyle, A. H. Trenkle and J. W. Young (1985). Changes in hormone and metabolite concentrations in plasma of steers during a prolonged fast. *J. Anim. Sci.* 61, 868-875.

SAS - Statistical Analysis Software Institute Inc. (2008). SAS/STAT® 9.2. User's Guide. Cary, NC, SAS Institute Inc.

Scheibe K. M. and W. J. Streich (2003). Annual rhythm of body weight in Przewalski horses (Equus ferus przewalskii). *Biol. Rhythm Res.* 34, 383-395.

Sjaastad O. V., K. Hove and O. Sand (2003). Physiology of Domestic Animals. Scandinavian Veterinary Press, Oslo, Norway.

Starinsky R. and E. Shafrir (1970). Displacement of albumin-bound bilirubin by free fatty acids. Implications for neonatal hyperbilirubinemia. *Clin. Chim. Acta* 29, 311-318.

Sticker L. S., D. L. Thompson, L. D. Bunting, J. M. Fernandez, C. L. DePew and M. R. Nadal (1995a). Feed deprivation of mares: Plasma metabolite and hormonal concentrations and responses to exercise. *J. Anim. Sci.* 73, 3696-3704.

Sticker L. S., D. L. Thompson, J. M. Fernandez, L. D. Bunting and C. L. Depew (1995b). Dietary-protein and(or) energy restriction in mares - plasma growth-hormone, ILGF-I, prolactin, cortisol, and thyroid-hormone responses to feeding, glucose, and epinephrine. *J. Anim. Sci.* 73, 1424-1432.

Suda A. K., C. S. Pittman, T. Shimizu and J. B. Chambers (1978). Production and metabolism of 3,5,3'-triiodothyronine and 3,3',5'-triiodothyronine in normal and fasting subjects. *J. Clin. Endocr. Metab.* 47, 1311-1319.

van Weyenberg S., M. Hesta, J. Buyse and G. P. J. Janssens (2008). The effect of weight loss by energy restriction on metabolic profile and glucose tolerance in ponies. *J. Anim. Physiol. An. N.* 92, 538-545.

von Engelhardt W. and G. Breves (2000). Physiologie der Haustiere. 3rd ed. Enke Verlag, Stuttgart, Germany.

Ward J. R., D. M. Henricks, T. C. Jenkins and W. C. Bridges (1992). Serum hormone and metabolite concentrations in fasted young bulls and steers. *Domest. Anim. Endocrin.* 9, 97-103.

Watson T. D. G., L. Burns, S. Love, C. J. Packard and J. Shepherd (1992). Plasma-lipids, lipoproteins and postheparin lipases in ponies with hyperlipemia. *Equine Vet. J.* 24:341-346.

Wintzer H-J. (1999). Krankheiten des Pferdes. 3rd ed. Paul Parey Verlag, Berlin, Germany.

Zeeb K. and U. Schnitzer (1997). Housing and training of horses according to their species-specific behaviour. *Livest. Prod. Sci.* 49, 181-189.

Znamirowksa A. (2005). Prediction of horse carcass composition using linear measurements. *Meat Sci.* 69, 567-560.

CHAPTER 4

Equilibration times of the isotopes ^{18}O, Deuterium and Tritium in animals studies. A meta analysis.

L. Brinkmann, M. Gerken

Department of Animals Sciences, University of Goettingen, Albrecht-Thaer Weg 3, 37075 Goettingen, Germany

Equilibration times of the isotopes ^{18}O, Deuterium and Tritium in animal studies. A meta analysis.

Lea Brinkmann, Martina Gerken

Department of Animals Sciences, Ecology of Livestock Production
Albrecht-Thaer Weg 3, 37075 Goettingen, Germany

Abstract

Isotopes are widely used as markers in animal studies. For these techniques the knowledge of the equilibration time (ET) of the isotopes is essential. 101 publications including 113 species belonging to birds, reptiles and mammals (ruminants, marsupials and monogastric animals) were reviewed. Data on ET were grouped into cohorts of Deuterium, Tritium and ^{18}O, body mass (BW), number of animals used, route of injection (ROI), amount of injected isotopes and medium (M) used for analysis. We additionally analysed the influence of the animal classes, the BW, the ROI and the M on the ET. While BW and animal classes had a significant influence on the ET of the isotopes Deuterium (P < 0.05), Tritium (P < 0.001) and ^{18}O (P<0.001), M was only significant for Tritium (P < 0.001) and ROI had no significant effect. The ET of nearly all animal classes differed significantly from each other. A prolonged ET was found for ruminants and reptiles. The ETs compiled in this study may serve as an orientation for further studies, but should be transferred with caution, considering possible impacts of the season, the reproductive status and differences between animal classes.

Keywords: FMR, equilibration time, isotope techniques, Deuterium, Tritium, ^{18}O

Introduction

The use of isotopes as markers in biological studies emerged around 1945 (Born, 1945; Fleischmann, 1945; Mattauch, 1947; Rabideau and Burr, 1945; Weiss et al., 1945). Since then the theoretical and methodological frame has been developed for several applications (Dziewiatkowski, 1951; Swindlehurst and Woodroofe, 1952; Lifson and Lee, 1959). Among the different isotopes, the stable non toxic isotopes ^{18}O, deuterated ($^{2}H_2O$) and the weak radioactive beta emitter tritiated water ($^{3}H_2O$, half-life 12 years) were mostly used as convenient means for measuring different physiological parameters in animal research such as the total body water, the water turnover rate (Nagy and Costa, 1980), the milk intake, the body composition and the energy expenditure (Midwood et al., 1993; Speakman et al., 2001; Fuller et al., 2004; Speakman and Krol, 2005; Riek, 2008).

Estimation of the energy expenditure in animals has often been conducted by measuring the respiratory gas exchange in metabolic chambers. This method only gives relative values of the energy costs of maintenance and controlled activity, but has little validity for the normal daily energy requirement of free ranging animals (Fuller et al., 2004). An alternative method to measure the energy expenditure of free living animals is the doubly labelled water method (DLW) which is based on the use of two different isotopes. The resulting energy expenditure derived from the DLW method is defined as field metabolic rate (FMR). The DLW was first described by Lifson et al. (1955), who performed subsequent studies on small rodents to validate the DLW method (McClintock and Lifson, 1958b; McClintock and Lifson, 1958a; McClintock and Lifson, 1958c; Schoeller and Vansanten, 1982). Many years passed before researchers began using the DLW method because of the very high cost for the isotopes needed to enrich the body water of subjects and the technical difficulty and relative unavailability of isotope measurements. Later deuterium was substituted by tritium, which was much easier to measure but large doses of ^{18}O were still required to yield accurate results. Hence isotope costs were still very high. Many studies were then performed on small animals but studies on large animals and humans remained impractical (Cole et al., 1990). The first human application was made in 1982 by Schoeller and Vansanten (1982). Today more than 200 publications exist on the FMR in animals, but still data from large animals are rare.

The estimation of the water turn over in the body and the body composition only requires one isotope, mostly deuterium or Tritium. The isotope dilution space can be estimated from the dilution of the isotope in the body over a given time and is used to assess the total body water content.

The DLW is based on the different turn over rates of two isotopes ^{18}O and Deuterium (^2H) respective Tritium (^3H) in the total body water (Fancy et al., 1986). Deuterium and Tritium leave the body water pool only through water losses like urine, sweat and evaporation whereas ^{18}O is additionally excreted via CO_2 (Fuller et al., 2004; Speakman, 2010). ^{18}O therefore leaves the body faster than Deuterium and Tritium. This results in a divergence in the isotope elimination rates in the body. By the knowledge of the respiratory quotient, the CO_2 production can be utilized to calculate the O_2 consumption and subsequently the FMR can be calculated (Riek et al., 2007b).

Isotopes can also be used to measure the milk transfer between mother and offspring. Two different methods have been developed. The first is based on the measurement of the decline of a known administered amount of hydrogen in the suckling offspring, which results from the dilution by ingested water (milk) in the suckling young. For this method milk needs to be the only source of water for the offspring. The second method uses isotope injection to the dam. The measurement is based on the transfer of isotopes via milk to the offspring and the simultaneous determination of the offspring's water turn over with another hydrogen isotope (Dove and Freer, 1979; Dove, 1988; Butte et al., 1991; Caire et al., 2002; Riek, 2006; Riek et al., 2007a).

The underlying assumption in the use of isotopes is the equal distribution of the isotopes in the body water pool after oral, intramuscular, intravenous or intraperitoneal application of the isotopes. Thus, the concentration of the isotopes in every compartment of the body is equal and any sample of body liquid (blood, urine, faeces, milk) is representative for the whole body. However, the isotopes take a certain time to distribute equally in the body compartments after the administration, the so defined equilibration time (ET). The isotope concentration at the ET is defined as the equilibration concentration. The ET can be most easily determined by taking several blood samples in short intervals after administrating the isotopes and analysing them for the isotope concentration. When the ET is reached, the isotope concentration begins to decline constantly.

The isotopes deuterium, tritium and ^{18}O are typically applicated as water isotopes. Although these isotopes come to equilibrium within the body rapidly because they appear to be very diffusible in water and the cell membrane (Coleman et al., 1972), the ET ranges from 15 minutes in small rodents like the common mouse (*Mus musculus*) (Krol and Speakman, 1999) to 14 hours in the Goanna (*Varanus caudolineatus*) (Thompson et al., 1997). The distribution of intravenous (iv) injected isotopes in all tissues dependents on the transport of isotopes within the blood circulation and the diffusion of isotopes through the interstitium to the intracellular space (Coleman et al., 1972). Thus, it is evident that the ET is longer in larger animals than in smaller ones because of the greater body mass in which they have to disperse.

Apart from body weight, several other factors may influence the ET such as, e.g., the route of injection, the type of isotopes and the source of body water used for analysis, as well as the metabolic rate of the animal (Speakman, 2010). The metabolic rate in turn depends among other things on ambient temperature, locomotor activity and growth rate of the animal (Gotaas et al., 2000). Thus, different seasons could influence the ET considerably. Furthermore, the ET in lactating and pregnant animals could differ from non-pregnant animals because of their changed metabolism, the particular water turn over and blood flow regarding the placenta and the embryo. There could be also a difference between the placentas of different species due to the varying blood permeability (Loeffler, 2002) so that the distribution of the isotopes may be delayed in same species.

The different isotopes (Deuterium, Tritium, ^{18}O) may have different dilution times even though they are administrated as water ($H_2^{18}O$, ^2H_2O and ^3H_2O) in most studies. There are five alternative routes for injection of the stable isotopes into the body: intravenous, intramuscular, intraperitoneal, subcutaneous (Speakman, 2010) and oral application (Andrews et al., 1997; Hinchcliff et al., 1997; Riek et al., 2007b). It is assumed that subcutaneous and especially oral application of the isotopes retard the time until equilibration occurs (Speakman, 2010). However, no divergence between intravenous application on different sites of the body and application into the aorta was found for the ET in dogs (Coleman et al., 1972). Degen et al. (1981) showed that the isotope concentration in blood samples after intramuscular and intravenous injection behave different during the period between injection and equilibration. After intravenous injection the isotope concentration in the blood raises rapidly in order to subsequently descent slowly until reaching equilibration. Intramuscular and intraperitoneal injection, however, tends to raising isotope concentration until equilibration (Denny and Dawson, 1975). Gotaas et al. (1997) explains this observation as follows: after administration the concentration of isotopes shows an initial peak in the respective body water compartment, followed by a fast decrease in the latter and at the same time an increase of concentration in the nonlabeled compartment. This fact should be kept in mind while applying isotopes and measuring the ET.

MacFarlane et al. (1969) indicates the possibility of a deviating dilution time in ruminant animals compared to monogastric animals, because of the large proportion (Farrell et al., 1972) and slow mixing of rumen water with the other body fluids (Macfarlane et al., 1969). Also the ET of deuterium and ^{18}O may diverge in ruminants due to the ^{2}H loss trough methane (Midwood et al., 1993).

Speakman (2010) advises as a 'rule of thumb' for the ET in animal studies to take one hour plus one hour for every 10 kgs of body mass. Exceeding a body weight of 100 kg the ET should be kept constant at about 10 hours. According to these recommendations this would result in an ET of one hour for a 20 g mouse and of 4 hours for a 30 kg dog and an Arctic walrus (*Odobenus rosmarus rosmarus*) with 1287 kg would have the same ET like the red deer (*Cervus elaphus*). However, the transfer of ETs from one animal species to another within the same weight category might be biased, because the ET might be affected by further variables other than body weight.

Despite of a large body of studies using isotopes, the information regarding the equilibration time or dosage of isotopes, which could be very useful in subsequent studies, is deficient. The aim of this study is to give an overview about the existing information and to analyse possible correlations between equilibration time and body mass. In addition the class of animals and the route of administration are considered.

Materials and Methods

Database

For this study publications were used in which ETs were given whether measured or estimated. If a time range was quoted the mean value was used.

This survey includes 101 published studies in which ^{18}O, Deuterium or Tritium were applied. In 29 % of the studies the authors measured the ET, in 20 % the ET was estimated but in 60.4 % it was not specified how the ET was determined. A total of 113 species of reptiles, marsupials, birds and mammals was included. The mammals were subdivided in ruminants, marsupials and monogastric animals because former studies showed that the FMR in ruminants and marsupials might diverge from that of other mammals (Midwood et al., 1989; Midwood et al., 1993; Riek, 2008).

ETs of Deuterium, Tritium and ^{18}O, data of the body mass and number of animals used are listed in Table 1. Furthermore, the route of administration, the amount of injected isotopes and the medium that was used for analysing the ET are given.

The animals were divided in birds, reptiles and mammals whereupon the latter were grouped into monogastric animals, marsupials and ruminants. Additionally the season and habitat in which the studies were conducted are listed. The Cohort includes: age of the animal, sex and reproductive status.

Statistical analysis

For the statistical analysis the following animal numbers were included for the respective isotopes: Deuterium: ruminants $n = 7$, birds $n = 11$, monogastric animals $n = 25$, reptiles $n = 4$; Tritium: ruminants $n = 10$, birds $n = 18$, monogastric animals $n = 19$, reptiles $n = 9$, marsupials $n = 28$; ^{18}O: ruminants $n = 9$, birds $n = 22$, monogastric animals $n = 43$, reptiles $n = 13$, marsupials $n = 17$.

All statistical analyses were performed with SAS, version 9.1.3 (SAS, 2009). For the analyses the means of the body weight (g) and the ET (min) of the respective species were used and $\log_{10}$ transformed. Publications lacking information on the BW were excluded from analysis. A linear regression of the ET on BW was made for Deuterium, ^{18}O and Tritium across all data and also for the different animal classes.

A 95% confidence interval (CI) of all data was calculated and data outside of the 95% CI were excluded from further analysis. The same procedure was applied for the data of each single animal class and after correction for outlayers the remaining data of the single animal classes were analysed together. MIXED procedures were performed for the dependent variable ET and the independent variables animal class, route of injection (Roi), medium used for analysis (M) and body weight for each isotope separately. Additionally the study was included as a random effect.

To compare the slopes of the ET at raising BW in the different animal classes a t-Test procedure was performed in which the difference of the slopes was divided by the standard error of the slopes.

Results

The analysed data (Tab.1) included very large animals like the Atlantic walrus *Odobenus rosmarus rosmarus* (1287 kg) as well as the tiny Marsupial honey possum *Tarsipes rostratus* (10 g). The mean BW of all analysed animals was 32.1 kg.

Within the studies revised, five publications did not specify the body weight of the studied animals and therefore could not be used for further regression analysis. Four publications were lacking information on the route of administration. In three studies the number of animals used was missing, 11 publications did not mention details on the amount of injected isotopes and in four publications the medium used for isotope analysis was not given.

Explicitly more studies used ^{18}O (N = 92) and Tritium (N = 79) than Deuterium (N = 37) for their analysis. In none of the Marsupial studies deuterium was used.

Various publications on the field metabolic rate and ET are lacking of information related to isotope application and information about the environmental conditions is spare. In most instances the isotope dosage is not clear as it is not explained if the injected amount was given per kg BW or for the entire animal and frequently the enrichment of the labels is not mentioned. Isotope concentrations, if presented, exhibit various units.

Tab. 1 Equilibration of mammals, reptiles and birds (adjusted for BW and order/suborder) together with route of injection, amount of injection, cohort, season, habitat, medium used for analyses and final blood sample (Chapter 4).

Class Order/ Suborder (*common name*)	n used in study	Mean BW (kg)	ROI[a]	Cohort[b]	Season[c]	Habitat[d]	Amount of injection (Abundance 100%) Deuterium (mL ^2H_2O) /kg BW	Tritium (μCi ^3H_2O) /kg BW	^{18}O (mL $H_2^{18}O$) /kg BW	M[e]	Equilibration time ^2H_2O (min)	^3H_2O (min)	$H_2^{18}O$ (min)	Final blood sample (h)	SOE[f]	Source
Mamalia (**Mammals**)																
Monogastric animals																
Mus domesticus (house mouse)	59	0.015	IP	M,F,BF	Su,A,W,Sp	Sc	-	-	6.49	a	-	90	90	48	N	Mutze *et al.* (1991)
Clethrionomys glareolus (Bank vole)	42	0.02	IP	A	-	L	0.45	-	0.9	a	60	-	60	24	E	Peacock *et al.* (2004)
Apodemus sylvaticus (Wood mouse)	19	0.021	IP	M, BM	-	SD, WL	0.34	-	0.77	a	90	-	90	48	N	Corp *et al.* (1999)
Gerbillus allenbyi (Allenby's gerbil)	43	0.023	IP	A	W,Su	D	-	770.93	7.49	a	-	120	120	168	E	Degen et al. (1992)
Gerbillus pyramidum (Greater egyptian gerbil)	29	0.033	IP	A	W,Su	D	-	536.32	5.21	a	-	120	120	168	E	Degen et al. (1992)
Mus musculus (Mouse)	3	0.037	IP	BF,F	-	L	0.28	1869.19	0.49	b	15	-	15	24	M	Król and Speakman (1999)
Acomys cahirinus (Common spiny mouse)	5	0.038	IP	A	Sp	D	-	29.61	3.83	a	-	120	120	192	E	Degen *et al.* (1986)
Sekeetamys calurus (Bushy tailed jird)	4	0.041	IP	A	Sp	D	-	27.44	3.55	a	-	120	120	192	E	Degen *et al.* (1986)
Microgale dobsoni (Dobson's shrew tenrec)	13	0.043	IP	A	Su	TF	0.94	-	1.28	a	90	-	90	48	E	Stephenson *et al.* (1994)
Microgale talazaci (Talazaci's long-tailed tenrec)	14	0.043	IP	A	Su	TF	0.93	-	1.28	a	90	-	90	48	E	Stephenson *et al.* (1994)
Pseudomys nanus (Barrow island mice)	40	0.044	IP	A	Sp, A	SS	-	50.0	0.29	a	-	120	120	168	N	Bradshaw *et al.* (1994)
Acomys russatus (Golden spiny mouse)	5	0.045	IP	A	Sp	D	-	25.0	3.23	a	-	120	120	192	E	Degen *et al.* (1986)

98

Tab. 1 continued.

Class Order/ Suborder (*common name*)	n used in study	Mean BW (kg)	ROI[a]	Cohort[b]	Season[c]	Habitat[d]	Amount of injection (Abundance 100%) Deuterium (mL ^2H_2O) /kg BW	Tritium (μCi ^3H_2O) /kg BW	^{18}O (mL $H_2{}^{18}O$) /kg BW	M[e]	Equilibration time ^2H_2O (min)	^3H_2O (min)	$H_2{}^{18}O$ (min)	Final blood sample (h)	SOE[f]	Source
Zyzomys argurus (Rock rat)	16	0.047	IP	A	Sp, A	SS	-	50.0	0.32	a	-	120	120	168	N	Bradshaw *et al.* (1994)
Microcebus murinus (Grey mouse lemur)	30	0.054	IP	A	D,W	DF	0.44	-	0.74	a	60	-	60	72	N	Schmid and Speakman (2000)
Ammospermophilus leucurus Antelope ground squirrels)	24	0.086	IP	A	Sp,Su,A	D	-	869.06	8.26	a	-	120	120	168	M	Karasov (1981)
Tamiasciurus hudsonicus (Red squirrel)	61	approx. 0.225	IP	F	W	CF	2.20	-	0.22	a	60	-	60	120	N	Humpries *et al.* (2005)
Spermophilus saturatus (Golden-mantled ground squirrel)	45	0.224	-	BA,J	Sp,Su,A	CF	-	1000.0	2.86	a	-	60	60	-	N	Kenagy et al. (1989)
Rattus (Rat)	21	0.227	IP	A	-	L	-	-	-	a	-	120	-	2	N	Culebras *et al.* (1977)
Rattus (Rat)	8	0.290	IP	M	-	L	0.06	-	-	a	120	-	120	120	M	Blanc *et al.* (2000)
Tamiasciurus hudsonicus (Red squirrel)	11	approx. 0.300	IP	A	Sp,A,W	CF	-	-	-	a	60	-	60	24	N	Bryce *et al.* (2001)
Sciurus carolinensis (Gray squirrel)	19	approx. 0.570	IP	A	Sp,A,W	CF	-	-	-	a	60	-	60	24	N	Bryce *et al.* (2001)
Xerus inauris (Cape ground squirrel)	18	0.578	IP	F	W	G	0.23	-	0.33	a	60	-	60	120	E	Scantlebury *et al.* (2007)
Suricata suricatta (Meerkats)	18	0.635	IP	F,BF	Sp,Su	D	0.25	-	0.11	a	60	-	60	168	M	Scantlebury *et al.* (2002)
Vulpes cana (Blanford foxes)	16	0.956	SC	A	Su,Wi	D	-	99.37	1.93	a	-	180	180	360	N	Geffen *et al.* (1992)
Martes americana (American Marten)	16	0.963	IP	A	A,W	DF	0.13	-	0.25	a	180	-	180	168	N	Gilbert *et al.* (2009)
Canis familiaris (Dog)	5	2.36	IV	J	-	-	0.32	-	0.67	a	150	-	150	-	M	Scantlebury *et al.* (2001)

99

Tab. 1 continued.

Class Order/ Suborder (*common name*)	n used in study	Mean BW (kg)	ROI[a]	Cohort[b]	Season[c]	Habitat[d]	Amount of injection (Abundance 100%) Deuterium (mL ^2H_2O) /kg BW	Tritium (µCi ^3H_2O) /kg BW	^{18}O (mL $H_2^{18}O$) /kg BW	M[e]	Equilibration time ^2H_2O (min)	^3H_2O (min)	$H_2^{18}O$ (min)	Final blood sample (h)	SOE[f]	Source
Marmota flaviventris (Yellow-bellied marmot)	35	2.99	IP	M	-	AM/SM	-	-	-	a	180	-	180	168	N	Salsbury and Armitage (1994)
Felis silvestris f. catus (Cat)	42	4.50	SC	A	-	L	0.19	-	0.02	a	240	-	240	336	N	Martin *et al.* (2001)
Sus scrofa domestica (Pig)	3	4.93	IP	J	-	L	0.17	-	0.32	b	30-40	-	30-40	-	M	Theil *et al.* (2007)
Vulpes vulpes (Red fox)	11	5.50	IM	A	Sp,A	T	-	0.91	0.18	a	-	180	180	-	N	Winstanley *et al.* (2003)
Alouatta palliata (Howler monkeys)	6	6.46	IV	A,J	D	TrF	-	147.06	2.02	a	-	180	180	264	E	Nagy and Milton (1979)
Proteles cristatus (Aardwolves)	6	8.16	IP	A	Su,W	SA	-	-	-	a	-	120	120	-	N	Williams *et al.* (1997)
Papio anubis (Olive baboon)	4	16.53	SC	BF	-	-	0.10	-	-	a	180 - 240	-	-	504	M	Garcia *et al.* (2004)
Canis familiaris (Dog)	5	18.4	-	A	-	-	-	32051.00	-	a	-	-	100	3.5	M	Coleman *et al.* (1972)
Canis familiaris (Dog)	8	24.3	oral	A	-	A	0.24	-	0.29	a	180	-	180	24	N	Hichcliff et al. (1997)
Lycaon pictus (African wild dog)	6	25.2	IV	A	-	D	0.12	-	0.06	-	90-180	-	90-180	96	M	Gorman *et al.* (1998)
Canis familiaris (Dog)	8	28.4	IV	A	-	L	0.07	-	0.13	a	300 - 360	-	300 - 360	23	M	Speakman *et al.* (2001)
Arctocephalus galapagoensis (Galapagos fur seals)	8	30.4	-	BF	-	M	0.07	-	0.14	a	90	-	90	288	M	Trillmich and Kooyman (2001)
Arctocephalus gazella (Antarctic fur seals)	21	32.1	IP	BF	Su	M	-	31.17	0.12	a	-	180	180	432	M	Costa *et al.* (1989)
Neophoca cinerea (Australien sea lion)	9	57.7	IP	J, A	Sp,W	M	-	26.0	0.18	a	-	180	180	192	N	Fowler *et al.* (2007)

Tab. 1 continued.

Class Order/ Suborder (*common name*)	n used in study	Mean BW (kg)	ROI[a]	Cohort[b]	Season[c]	Habitat[d]	Amount of injection (Abundance 100%)			M[e]	Equilibration time			Final blood sample (h)	SOE[f]	Source
							Deuterium (mL ^2H_2O) /kg BW	Tritium (μCi ^3H_2O) /kg BW	^{18}O (mL $H_2^{18}O$) /kg BW		^2H_2O (min)	^3H_2O (min)	$H_2^{18}O$ (min)			
Neophoca cinerea (Australien sea lion)	20	69.25	IP	BF	Su.W	M	-	0.86	0.15	a	-	180	180	-	N	Costa and Gales (2003)
Phocarctos hookeri (New zealand sea lions)	12	113.9	IP	BF	Su	M	-	42736.00	0.12	a	-	180 - 240	180 - 240	240	N	Costa and Gales (2000)
Equus ferus caballus (Horse)	3	approx. 265.0	IV	M	-	L	0.83	-	0.04	a	180-300	-	180-240	336	M	Fuller *et al.* (2004)
Equus ferus caballus (Horse)	6	503.4	oral	A	-	-	0.14	-	-	a	120 - 180	-	-	24	M	Andrews *et al.* (1997)
Odobenus rosmarus rosmarus (Atlantic walrus)	6	1287.0	IV	A	Su	M	0.07	-	-	a	120 - 168	-	-	456	M	Acquarone and Born (2007)
Oryctolagus cuniculus (Rabbit)	2	-	IV	A	-	-	-	-	-	-	60 - 90	60 - 90	60 - 90	2.3	M	Anbar and Lewitus (1958)
Halichoerus grypus (Grey seals)	9	-	IV	F	-	M	-	-	-	a	180	-	180	120	N	Sparling *et al.* (2008)
Ruminantia (*Ruminants*)																
Odocoileus hemionus columbianus (Black tailed deer)	5	45.4	IV	A	Sp	TM	0.10	-	0.10	a	300-480	-	300-480	312	N	Nagy *et al.* (1990)
Ovis orientalis aries (Sheep)	24	45.6	IM	BF	A	TM	-	27120.00	-	milk	-	360	-	-	N	Dove (1988)
Lama pacos (Alpaca)	4	48.0	oral	M	Su	Sa	-	-	0.01	a	-	-	360	336	N	Riek *et al.* (2007b)
Lama pacos (Alpaca)	16	48.7	oral	M	Su	Sa	0.14	-	-	a	360	-	-	336	N	Riek *et al.* (2007b)
Capra aegagrus f. hircus (Angora goat)	8	51.0	IV	M	-	-	-	-	-	a	360	-	360	672	N	Toerien *et al.* (1999)
Bibos banteng (Bali cattle)	2	51.0	IM	J	-	TM	-	10.00	-	a	-	120	-	360	M	Macfarlane *et al.* (1969)

Tab. 1 continued.

Class Order/ Suborder (*common name*)	n used in study	Mean BW (kg)	ROI[a]	Cohort[b]	Season[c]	Habitat[d]	Amount of injection (Abundance 100%)			M[e]	Equilibration time			Final blood sample (h)	SOE[f]	Source
							Deuterium (mL ^2H_2O) /kg BW	Tritium (μCi ^3H_2O) /kg BW	^{18}O (mL $H_2{}^{18}O$) /kg BW		^2H_2O (min)	^3H_2O (min)	$H_2{}^{18}O$ (min)			
Sminthopsis crassicaudata (Fat - tailed dunnart)	15	0.015	IP	A,J	Sp	TM	-	2200.00	6.33	a	-	60	60	48	E	Nagy *et al.* (1988)
Phascogale calura (Wambenger)	51	0.034	IP	A	W,Su,Sp	EF	-	15882.35	2.79	a	-	120 - 180	120 - 180	96	N	Green *et al.* (1989)
Petaurus breviceps (Sugar glider)	36	0.120	IP	A	Sp,Su,A	EF	-	2702.50	3.17	a	-	120-180	120-180	120	E	Quin *et al.* (2010)
Gymnobelideus leadbeateri (Leadbeater´s Possum)	6	0.121	IP	A	W	EF	-	3526.97	-	a	-	180	-	192	N	Smith *et al.* (1982)
Gymnobelideus leadbeateri (Leadbeater´s Possum)	18	0.129	IP	A	Sp	EF	-	466.93	0.99	a	-	180	-	120	N	Smith *et al.* (1982)
Isoodon auratus (Golden bandicoot)	60	0.333	IP	A	Sp,A	SS	-	500.00	2.91	a	-	120	120	168	N	Bradshaw *et al.* (1994)
Isoodon auratus (Golden bandicoot)	8	0.358	IM	A	Su	SS	-	1000.00	0.63	a	-	120	120	72	N	Nagy and Bradshaw (2000)
Bettongia lesueur (Burrowing bettong/ Boodie)	6	0.720	IM	A	Su	SS	-	100.00	0.15	a	-	180	180	96	N	Nagy and Bradshaw (2000)
Potorous tridactylus (Long nosed-potoroo)	20	0.800	IM	A	Sp, Su	EF	-	125.00	0.34	a	-	180	180	288	N	Wallis *et al.* (1997)
Pseudocheirus peregrinus (Common ringtail possum)	100	0.996	IP	M,F,BF	Su,A,W,Sp	EF	-	5018.07	0.29	a	-	180	180	144	E	Munks and Green (1995)
Trichosurus arnhemensis (Northern brush possum)	7	1.10	IM	A	Su	SS	-	100.00	0.15	a	-	180	180	144	N	Nagy and Bradshaw (2000)
Isoodon obesulus (Short-nosed bondicoot)	10	1.23	IP	A	A	EF	-	990.00	2.85	a	-	120	120	144	N	Nagy *et al.* (1991)
Potorous tridactylus tridactylus (Potoroo)	4	1.50	IP	A	-	SF	-	150.00	-	a	-	240	-	-	M	Denny and Dawson (1975)
Petrogale lateralis (Werstern rock-wallaby)	1	2.21	IM	A	Su	SS	-	75.01	0.15	a	-	180	180	144	N	Nagy and Bradshaw (2000)

102

Tab. 1 continued.

Class Order/ Suborder (*common name*)	n used in study	Mean BW (kg)	ROI[a]	Cohort[b]	Season[c]	Habitat[d]	Amount of injection (Abundance 100%)			M[e]	Equilibration time			Final blood sample (h)	SOE[f]	Source
							Deuterium (mL ^2H_2O) /kg BW	Tritium (μCi ^3H_2O) /kg BW	^{18}O (mL $H_2{}^{18}O$) /kg BW		^2H_2O (min)	^3H_2O (min)	$H_2{}^{18}O$ (min)			
Lagorchestes conspicillatus (Spectacled hare-wallaby)	5	2.53	IM	A	Su	SS	-	75.01	0.15	a	-	180	180	144	N	Nagy and Bradshaw (2000)
Thylogale billardierii (Tasmanian pademelon)	5	5.98	IV	A	Su	EF	-	830.94	2.37	a	-	240	240	264	N	Nagy *et al.* (1990)
Macropus eugenii (Tammar wallaby)	6	6.60	IP	A	-	SF	-	150.00	-	a	-	240	-	-	M	Denny and Dawson (1975)
Phascolarctos cinereus (Koala)	3	approx. 7.5	IV	A	-	EF	-	64.87	-	a	-	120	-	-	M	Ellis *et al.* (1995)
Phascolarctos cinereus (Koala)	6	9.20	IV	A	Sp	EF	-	600.00	2.87	a	-	120	120	456	E	Nagy and Martin (1985)
Macropus robustus isabellinus (Barrow island euro)	5	12.91	IV	A	Su	SS	-	75.00	0.15	a	-	180	180	192	N	Nagy and Bradshaw (2000)
Macropus giganteus (Grey kangaroo)	2	16.30	IP	A	-	EF	-	61.35	-	a	-	300 - 360	-	12	M	Denny and Dawson (1975)
Macropus giganteus (Grey kangaroo)	1	17.20	IV	A	-	EF	-	58.14	-	a	-	330	-	12	M	Denny and Dawson (1975)
Megaleia rufa (Red kangaroo)	1	21.40	IV	A	-	G	-	46.73	-	a	-	300	-	12	M	Denny and Dawson (1975)
Megaleia rufa (Red kangaroo)	1	21.90	IP	A	-	G	-	45.66	-	a	-	240	-	12	M	Denny and Dawson (1975)
Macropus robustus robustus (Wallaroo)	6	31.10	IP	A	-	EF	-	24.98	-	a	-	360	-	-	M	Denny and Dawson (1975)
Marsupialia Vombatidae (Wombat)	3	approx. 33.0	IP	F	-	-	-	151.52	-	a	-	240	-	24	M	Evans *et al.* (2003)
Macropus giganteus (Grey kangaroo)	2	43.90	IV	M	Su	EF	-	36.51	0.10	a	-	240	240	384	N	Nagy *et al.* (1990)

Tab. 1 continued.

Class Order/ Suborder (common name)	n used in study	Mean BW (kg)	ROI[a]	Cohort[b]	Season[c]	Habitat[d]	Amount of injection (Abundance 100%) Deuterium (mL ^2H_2O) /kg BW	Tritium (µCi ^3H_2O) /kg BW	^{18}O (mL $H_2^{18}O$) /kg BW	M[e]	Equilibration time ^2H_2O (min)	^3H_2O (min)	$H_2^{18}O$ (min)	Final blood sample (h)	SOE[f]	Source
Reptilia (Reptiles)																
Varanus caudolineatus (Goanna)	9	0.010	IP	A	Su	Sa, Sc	-	2403.85	5.67	a	-	600	600	140	N	Thompson *et al.* (1997)
Varanus caudolineatus (Goanna)	10	0.014	IP	A	-	L	-	1760.56	4.15	a	-	840	840	140	N	Thompson *et al.* (1997)
Oedura marmorata (Marbled velvet gecko)	39	0.015	IP	A	Sp,Su,W,	Tr, T	-	54000.00	1.60	a	-	360	360	-	N	Christian *et al.* (1998)
Agama impalearis (Bibrons agama)	12	0.054	IP	A	Su	D	-	6111.11	4.31	a	-	120	120	360	E	Znari and Nagy (1997)
Boiga irregularis (Brown treesnake)	14	0.132	IP	A	D,W	TF	0.59	-	0.54	a	120	-	120	192	N	Anderson *et al.* (2003)
Lophognathus temporalis (tropical dragon)	31	0.366	IP	A	D,W	Tr	-	1108.07	0.97	a	-	480	480	984	N	Christian *et al.* (1999)
Tiliqua scincoides (Bluetongue lizards)	15	0.576	-	A	D,W	Tr	-	-	-	a	-	480 - 720	480 - 720	1296	N	Christian *et al.* (2003)
Chelodina longicollis (Freshwater turtle)	19	0.612	IP	A	Su	W	-	1633.90	0.39	-	-	240 - 300	240 - 300	-	N	Roe *et al.* (2008)
Iguana iguana (Green iguana)	3	approx. 0.860	IP	A	-	L	-	-	-	a	720	-	720	-	M	van Marken Lichtenbelt *et al.* (1993)
Varanus rosenbergi (Rosenberg's goanna)	44	1.17	IP	A	W,Sp,Su	EW	-	4277.16	0.28	a	-	360	360	2880	N	Green *et al.* (1991a)
Amblyrhynchus cristatus (Marine iguana)	38	approx. 2.18	IP	M	-	M	0.25	-	0.45	a	480 - 600	-	480 - 600	192	N	Drent *et al.* (1999)
Varanus gianteus (Perenties)	32	5.57	IP	A	A,Su	SS	-	873.43	0.17	a	-	360	360	936	N	Green *et al.* (1986)
Chelonia mydas L. (Green turtle)	6	22.42	IC	A	-	L	-	-	-	a	300	-	300	72	M	Jones *et al.* (2009)

104

Class Order/ Suborder (*common name*)	n used in study	Mean BW (kg)	ROI[a]	Cohort[b]	Season[c]	Habitat[d]	Amount of injection (Abundance 100%) Deuterium (mL ^2H_2O) /kg BW	Tritium (µCi ^3H_2O) /kg BW	^{18}O (mL $H_2^{18}O$) /kg BW	M[e]	Equilibration time ^2H_2O (min)	^3H_2O (min)	$H_2^{18}O$ (min)	Final blood sample (h)	SOE[f]	Source
Varanus komodoensis (Komodo dragon)	1	45.20	IP	A	Sp,Su	Tr	-	110.62	0.05	a	-	over night	over night	240	N	Green *et al.* (1991b)

Aves (Birds)

Class Order/ Suborder (*common name*)	n used in study	Mean BW (kg)	ROI[a]	Cohort[b]	Season[c]	Habitat[d]	Deuterium (mL ^2H_2O) /kg BW	Tritium (µCi ^3H_2O) /kg BW	^{18}O (mL $H_2^{18}O$) /kg BW	M[e]	^2H_2O (min)	^3H_2O (min)	$H_2^{18}O$ (min)	Final blood sample (h)	SOE[f]	Source
Estrilda Troglodytes (Black-rumped waxbills)	9	0.007	IM	A	-	L	-	223881.00	5.67	a	-	60	60	24	N	Weathers and Nagy (1984)
Troglodytes aedon (House wren)	17	0.010	IM	J	-	DF	-	6250.00	5.94	a	-	60	60	28	E	Dykstra and Karasov (1993)
Acanthorhynchus tenuirostris (Eastern spinebill)	6	0.010	IM	A	S	EF	-	2226.80	5.98	a	-	60	60	48	N	Weathers *et al.* (1996)
Eremiornis carteri (Spinifexbird)	47	0.012	IM	A	-	SS	-	8333.33	11.67	a	-	180	180	30	N	Ambrose *et al.* (1996)
Phylidonyris pyrrhoptera (Crescent honeyeater)	1	0.015	IM	A	S	EF	-	1479.45	3.97	a	-	60	60	48	N	Weathers *et al.* (1996)
Phylidonyris novaehollandiae (New holland honeyeater)	12	0.017	IM	A	S	EF	-	1270.59	3.41	a	-	60	60	48	N	Weathers *et al.* (1996)
Junko hyemalis (Dark-eyed junco)	24	0.020	IM	M	Sp,Su	L	-	800.00	2.43	a	-	60	60	24	N	Lynn *et al.* (2000)
Lanius ludovicianus (Loggerhead shrike)	8	0.046	IM	A	S	D	-	2197.80	5.23	a	-	60	60	24	N	Weathers at al. (1984)
Sturnus roseus (Rose coloured starling)	8	0.071	IP	A	-	L	0.99	-	1.69	-	60	-	60	144	N	Engel *et al.* (2006)
Turdoides squamiceps (Arabian babbler)	84	0.073	oral	A,BF,BM	W,Su,Sp	D	-	1721.76	0.66	d	-	45 - 60	60	48	E	Anava *et al.* (2002)
Calidris canutus (Red Knot)	6	0.108	SC	A	-	L	-	-	-	a	60	-	-	-	E	Visser et al. (2000)
Coturnix c. japonica (Japanese quail)	9	0.254	IP	M	-	L	1.50	-	1.38	a	60	-	60	24	E	Van Trigt *et al.* (2002)

Tab. 1 continued.

Class / Order/ Suborder (*common name*)	n used in study	Mean BW (kg)	ROI[a]	Cohort[b]	Season[c]	Habitat[d]	Deuterium (mL ^2H_2O) /kg BW	Tritium (µCi ^3H_2O) /kg BW	^{18}O (mL $H_2^{18}O$) /kg BW	M[e]	^2H_2O (min)	^3H_2O (min)	$H_2^{18}O$ (min)	Final blood sample (h)	SOE[f]	Source
Alectoris chukar (Chukar partridges)	3	0.365	IV	A	A	D	-	54.79	-	a	-	45	-	3	M	Degen *et al.* (1981)
Rissa tridactyla (Kittiwakes)	53	0.366	IM	A	Su	M	-	1092.90	2.86	a	-	60	60	72	E	Fyhn *et al.* 2001
Rissa tridactyla (Kittiwakes)	67	0.376	IM	BA	-	M	0.66	-	0.99	a	60		60	72	E	Welker *et al.* 2010
Cepphus grylle (Black guillemot)	9	0.380	IM	BF,BM	Su	M	-	1252.63	3.04	a	-	60	60	72	E	Mehlum et al. (1993)
Alectoris chukar (Chukar partridges)	3	0.392	IM	A	A	D	-	51.02	-	a	-	25	-	3	M	Degen *et al.* (1981)
Gallus gallus domesticus (Chicken)	1	1.05	IP	A	-	L	0.19	950.57	0.36	a	60	60	60	96	N	Gessaman *et al.* (2004)
Eudyptula minor (Little penguin)	-	approx. 1.20	IP	BA,M,F	Sp,Su,A,W	M	-	4166.67	0.28	a	-	300 - 3600	300 - 360	336	M	Gales and Green (1990)
Somateria mollissima (Eider duck)	6	1.79	IM	A	-	L	0.15	-	0.27	a	240	-	240	48	N	Hawkins *et al.* (2000)
Branta leucopsis (Barnacle goose)	21	2.00	IP	A	-	TM	0.56	-	-	a	90	-	-	72	M	Eichhorn and Visser (2008)
Morus capensis (Cape gannet)	21	2.58	IM	BA	-	CS	0.09	-	0.09	a	120	-	120	-	N	Adams *et al.* (1991)
Porphyrio mantelli (Takahe)	6	2.58	IP	A	A	TeF	-	-	-	a	105	105	105	72	N	Godfrey *et al.* (2003)
Macronectes giganteus (Southern giant-petrels)	8	3.89	IM	BA	Su	A	-	534.45	0.33	a	-	150	150	-	N	Obst and Nagy (1992)
Pygoscelis adeliae (Adelie penguins)	48	3.97	IM	BA	Su	A	0.25	-	0.14	a	120 - 180	-	120 - 180	-	N	Chappell *et al.* (1993)

106

Tab. 1 continued.

Class Order/ Suborder (*common name*)	n used in study	Mean BW (kg)	ROI [a]	Cohort[b]	Season[c]	Habitat[d]	Amount of injection (Abundance 100%)			M[e]	Equilibration time			Final blood sampl e (h)	SOE[f]	Source
							Deuterium (mL ^2H_2O) /kg BW	Tritium (μCi ^3H_2O) /kg BW	^{18}O (mL $H_2^{18}O$) /kg BW		^2H_2O (min)	^3H_2O (min)	$H_2^{18}O$ (min)			
Aptenodytes patagonicus (King penguin)	-	12.9	IP/IM	A	A	M	0.18	-	0.07	a	180	-	180	216	N	Kooyman et al. (1992)
Struthio camelus (Ostrich)	1	114.0	IV	A	-	D	-	87.72	-	a	-	120	-	-	N	Williams (1993)

[a]Route of injection: IV, intravenous; IM, intramuscular; IP, intraperitoneal; SC, subcutan; R, injektion through the body wall into rumen

[b]Cohort: A , adult; M, male; F, female; BF, breeding/lactating female; BM, breeding male; J, juvenile

[c]Season: W, wet; D, dry; Wi, winter; Su, summer; Au, autum; Sp, spring

[d]Habitat: L = Lab; A, antarctica/arctic; Tr, tropical; Cs, chaparral scrub; D, desert; SS, semiarid scrub; TF, tropical forest; EF, euclypt forest; M, marine; G, grassland; sf, sclerophyll forest; EW, eucalyptus woodland

DF, deciduous forest; TM, temerate meadow; T, temperate; Sa, semi arid; TeF, temperate forest; P, pasture; AM/SM, alpine-subalpine meadow; TT, taiga/ tundra; W, wetland; cf, coniferous forest; lab, laboratory/aviary;

[e]medium used for analysis: a, blood; b, breath; c, urine; d, feaces

[f]Source of ET: M, measured; N, method of ET determination not given; E, estimated or transfered

Fig. 1-3 depict all data for Deuterium, Tritium and ^{18}O from table 1 in a common Y versus X graph. The plots show a linear regression line without including the study effect.

All tree figures indicate that the ET increase with rising body weight. The inclination for Deuterium (ET ^{2}H$_2$O, minutes = 1.51 + 0.18 log$_{10}$ BW, g) is significantly higher (p < 0.01) than for tritium (ET ^{3}H$_2$O, minutes = 1.95 + 0.08 log$_{10}$ BW, g). The slope of ^{18}O (ET H$_2$^{18}O, min = 1.81 + 0.12 log$_{10}$ BW, g) does not differ significantly from those on tritium and deuterium, but the ET of Deuterium at low BW is shorter than in Tritium and ^{18}O

The dispersion of the ETs is relatively high and the ET can have different durations at the same body weight. Only 29% of the ET variation in deuterium, 13% in Tritium and 21% in ^{18}O is explained by the variation in BW.

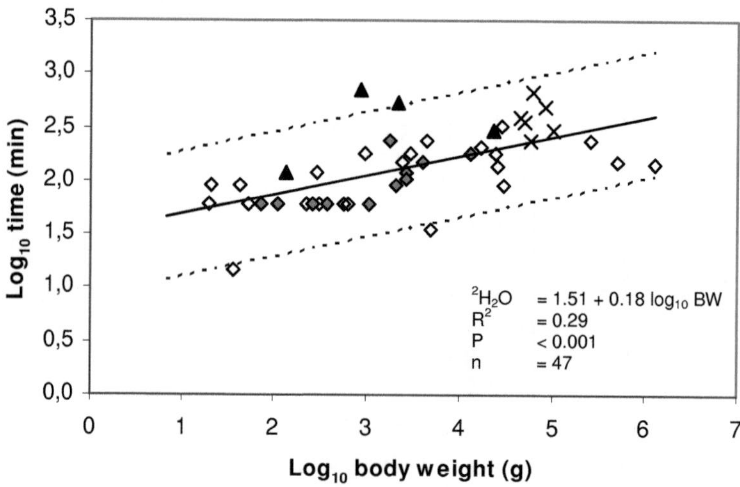

Fig. 1 Relationship between mean deuterium ET and mean body mass in 42 different animal species (47 single animal data from 44 published studies). Plot shows a linear regression line without study effect. The 95% confidence interval of the prediction of all data is displayed by the dashed lines. (rhomb = monogastric animals, cross = ruminants, coloured rhomb = birds, coloured triangle = reptiles).

No difference could be detected between the results of the analysis of variance containing the data within the 95% CI of all animals and the analysis of variance of the respective animal classes. Therefore, only the results of the variance analysis containing the data within the 95% CI of all animals are shown in Table 2.

In both analyses the ET is significantly affected by the BW (Deuterium $p < 0.05$; Tritium $p < 0.001$; ^{18}O $p < 0.001$) and the animal class (Deuterium $p < 0.05$; Tritium $p < 0.001$; ^{18}O $p < 0.001$) even though the study is included as a random effect (Table 2). A significant influence of the medium used for analysis on the ET could be only observed for Tritium ($p < 0.001$). There is hardly any difference between the significance levels of analyses with or without the study as a random effect. No effect on the ET of the isotopes could be detected regarding the route of injection. The t-tests revealed no significant differences in the slopes of ET for the different animal classes.

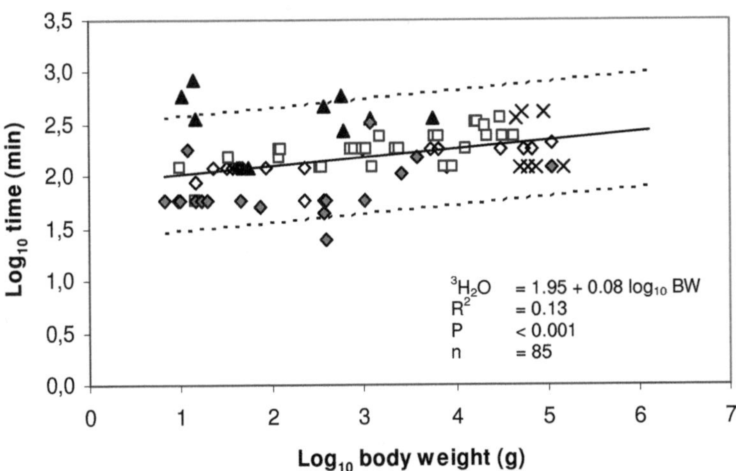

Fig. 2 Relationship between mean tritium ET and mean body mass in 75 different animal species (85 single animal data from 60 published studies). Plot shows a regression line without study effect. The 95% confidence interval of the prediction of all data is displayed by the dashed lines (rhomb= monogastric animals, cross = ruminants, coloured rhomb = birds, coloured triangle = reptiles, square = marsupials).

Table 3 shows the differences in the ET (BW corrected) between the five animal classes for every isotope. Presented are both the results of the data excluding the outliers (information outside the 95% CI of prediction) of all animals and the results excluding the outliers (data outside the 95% CI of prediction) of every single animal class. Both analyses differ only in few animal class comparisons. Deuterium was not used for marsupial analysis and is therefore not included in the table. In the analysis of all animal data, 17 animal class comparisons are significantly different, in the analysis of the single animal data 18 animal class comparisons differ significantly. All animal classes differ at least in one isotope.

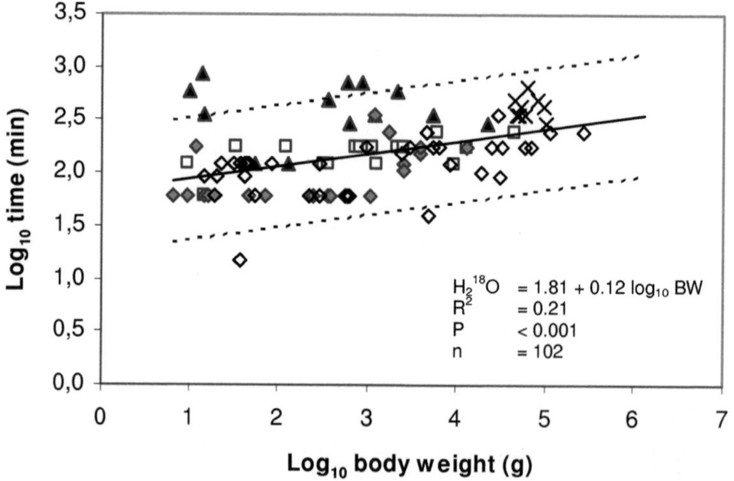

Fig. 3 Relationship between the mean ET of heavy water and mean body mass in 94 different animal species (102 single animal data from 88 published studies). Plot shows a regression line without study effect. The 95% confidence interval of the prediction of all data is displayed by the upper and lower lines (rhomb = monogastric animals, cross = ruminants, coloured rhomb = birds, coloured triangle = reptiles, square = marsupials).

The comparison of the slopes of the ET on BW between different animal classes showed no significant differences since there were too little data for each of them. In summary, there were the following tendencies: for both, the data excluding the outliers (outside the 95% CI of prediction) of all data and the data excluding the outliers of the respective animal classes, the ascending slopes of the monogastric animals and the birds were similar. Ruminant and reptile slopes clearly diverged from the other animal classes and show almost no increase. The marsupials slope is not varying much from the one of monogastric animals but crosses the letter, so that there is a higher ET for small and a lower ET for larger marsupials compared with monogastric animals.

Tab. 2 Results of the variance analysis containing the data within the 95% CI of prediction of all animals. Effect of the variables BW (log_{10}), animal class, medium used for analysis (M) and route of injection (RoI) on the ET of Deuterium (^2H_2O), Tritium (^3H_2O) and heavy water ($H_2^{18}O$).

Equilibration time						
	^2H_2O		^{18}O		^3H_2O	
	F	p	F	p	F	p
Simple Varianz Analysis						
BW (log_{10})	6,13	0,020	33,95	<0.001	31,1	<0.001
Species	3,72	0,023	10,51	<0.001	13,6	<0.001
M	0,77	n.s.	0,07	n.s.	11,38	<0.001
RoI	1,71	n.s.	0,81	n.s.	0,88	n.s.
Study included as random effect						
BW (log_{10})	6,13	0,020	33,95	<0.001	31,1	<0.001
Species	3,72	0,024	10,51	<0.001	13,6	<0.001
M	0,77	n.s.	0,07	n.s.	11,38	<0.001
RoI	1,71	n.s.	0,81	n.s.	0,88	n.s.

Tab. 3 Significance (p) of the slope comparisons of ET on BW for Deuterium (^2H_2O), Tritium (^3H_2O) and ^{18}O between marsupials (ma), ruminants (ru), monogastric animals (mo), birds (b) and reptiles (rep) after excluding data outside the 95% CI for all animal data and for data of the particular animal class respectively. Deuterium was not used in marsupial studies.

95% CI of prediction						
	of all data			of the respective animal class		
Animal class	^2H_2O	^3H_2O	^{18}O	^2H_2O	^3H_2O	^{18}O
mo vs. ru	0.016	0.025	0.008	0.028	n.s.	0.011
mo vs. ma	-	0.039	n.s.	-	n.s.	n.s.
mo vs. rep	n.s.	< 0.001	< 0.001	< 0.001	< 0.001	< 0.001
mo vs. b	n.s.	0.014	n.s.	n.s.	< 0.001	n.s.
ru vs. ma	-	< 0.001	n.s.	-	0.003	n.s.
ru vs. rep	n.s.	< 0.001	n.s.	n.s.	< 0.001	0.022
ru vs. b	0.004	n.s.	< 0.001	0.008	n.s.	< 0.001
ma vs. rep	-	< 0.001	< 0.001	-	< 0.001	< 0.001
ma vs. b	-	< 0.001	0.004	-	< 0.001	0.007
rep vs. b	n.s.	< 0.001	< 0.001	< 0.001	< 0.001	< 0.001

Discussion

The aim of this study was to give an overview on existing equilibration time data for the isotopes ^{18}O, Deuterium and Tritium in animal studies and to identify influencing effects.

We used data from 113 species but only five species used weighted over 100 kg. This might be explained by the fact that isotopes are still expensive and big animals are more difficult to handle and to house. There were other studies published on the FMR of large animals, but ETs or other necessary data like the BW were not mentioned, so that this information could not be included in our present analysis.

The data on the applied isotopes were often insufficient. The amount of isotopes injected in the animals deviated highly and we noted a tendency that smaller animals receive a higher amount of isotopes per kg BW then large animals. The injected amount of isotopes may be reduced to a minimum in large animals to keep the costs down, but this can also be caused by the fact that big animals were mostly used later in time when the isotopes became cheaper. This was achieved by improved analysis techniques to detect lower isotope concentrations in the media and hence fewer amounts of isotope applications were needed.

The data were arranged into the animal classes birds, reptiles and mammals. The animal class of mammals was subdivided into ruminants, marsupials and monogastric animals because marsupials were claimed to have a lower metabolic rate than eutherians (Tyndale-Biscoe, 2005) and ruminants were presumed to differ from monogastric animals because of their gastrointestinal system (see below). The class of the ruminants in this study consists of true ruminants as well as camelids, even though camelids do not actually belong to the true ruminants (Fowler, 2008). However, the camelids fit best in this group, being ruminating animals and having a comparable FMR (Riek et al., 2007b).

The animal class influences the ET significantly ($P < 0.001$) with ruminants and reptiles taking a longer time until the isotopes are equilibrated equally in the body water. The BW corrected ETs of these animal classes differ nearly for all isotopes from those in birds, marsupials and monogastric animals. Midwood et al. (1993) noted that the large amount of water in the gastrointestinal system in ruminants can have an influence on the rate of equilibration of isotopically labelled water. The slower mixing of rumen water with the other body fluids apparently extends the dilution time in ruminant animals (Macfarlane et al., 1969). In addition Williams (2001) pointed out that ruminants have a lower metabolic rate than Artiodactyla which may also influence the ET.

Reptiles show a lower FMR than most marsupials, mammals and ruminants (Dryden et al., 1990; Green et al., 1991a; Christian et al., 1999; Christian et al., 2003) and their longer ET may result from this. Similarly large Marsupials have a slower basal metabolic rate of about 30 – 35% than other similar sized eutherians. In contrast, small marsupials show a higher metabolic rate than eutherians of the same size (Bradshaw and Bradshaw, 2007; Riek, 2008). This may explain the significant ET differences compared to the other animal classes and coincides with our observations of the crossing slopes in marsupials and monogastric animals.

Calculating separate regression equations of ET on BW for the various animal classes after excluding outliers did not lead to significant differences between animal classes in the slopes of the regression lines (*t*-test, $P > 0.05$). This may be explained by the low number of studies included in either animal class. The regression of the ruminants and the reptiles showed no significant inclination with raising BW.

As expected, larger animals had a longer time until isotopes were equally diluted in the body water, shown by the significant influence of the BW ($p < 0.001$) on the ET.

The present increase of the ET with rising BW is not as high as Speakman (2010) suggested in his rule of thumb. When Spearman's assumptions are log_{10} transformed, the increase would be 0.7 units of ET per BW unit. In contrast, in this study the increase lies between 0.08 and 0.18 units. The variation in BW only accounts for 12% to 29% of the variation in ET depending on the isotope and varying ETs exist for similar body weights (Fig. 1-3). This large spread around the regression line can be caused by various effects, such as the route of isotope injection. Our analysis, however, could not detect significant influences of the route of injection on the ET. Denny and Dawson (1975) as well as Degen et al. (1981) confirm our results, which did not discover different ETs after *iv* or *ip* injection of isotopes in macropodid marsupials or chukar partridges, respectively. Contrary to this observations Smith and Sykes (1974) found a more rapid equilibration after intravenous than after intraperitoneal injections and Speakman (2010) pointed out that oral application retard the time until equilibration occurs. This difference in ET due to the route of injection can be ascribed to the fact that the isotopes disperse mainly by diffusion through the interstitium and especially after intravenous application the distribution takes a shorter time because the isotopes distribute very fast through the blood and then through the interstitium (Coleman et al., 1972). In the present study no difference between oral and other applications was revealed because of the small data volume regarding orally applied isotopes.

The ET of the isotopes might also be influenced by the isotope applied. In most of the reviewed studies, like e.g. Anbar und Lewitus (1958) in monogastric animals, Toerien et al. (1999) in ruminants, Nagy and Bradshaw (2000) in marsupials, Christian et al. (1999) in reptiles and Weathers et al. (1996) in birds, no variation of ET for the isotopes ^{18}O, Deuterium and Tritium was described. Only Fuller et al. (2004) and

Anava et al. (2002) discovered different ET`s for ^{18}O and Deuterium in horses and for ^{18}O and Tritium in Arabian babblers, respectively. Varying ETs for different isotopes may be attributed to the fact that for large animals the two pool model is used to measure the FMR because hydrogen has a larger dilution space (Speakman and Krol, 2005) than oxygen. The ET of oxygen may therefore be shorter than that of hydrogen. A tendency for differences between the dilution velocity of the isotopes is indicated by the different slopes of the regression lines, although only tritium and deuterium differ significantly. However, a significant difference would have been more expected between ^{18}O and the hydrogen isotopes rather than between hydrogen isotopes as they should behave similar in the body water because of their chemical structure. The observed difference in slopes may result from the primarily use of tritium in marsupials, in which deuterium was not applied.

Differences in the FMR may also influence the time until the isotopes have equally distributed within the body water and would be faster with a higher metabolic rate. Accordingly, ET would be influenced by the season (e.g. ambient temperature, water availability), the age cohort, the habitat, the physiological status (e.g. stress, lactating, pregnant) and the sex of the animal. Data of the age cohort, season and habitat could not be considered in our statistical analysis because effects of seasons and age cohorts were intermingled and the habitats were different for nearly every study or information were lacking. The dry season of the year can lead to dehydration in animals and can accordingly lengthen the ET (Siebert and MacFarlane, 1971). It is well established that the metabolic rate of mammals is highly correlated with their body temperature (e.g. hypothermia and hibernation) (Schmidt-Nielsen et al., 1967) so that the season can influence the ET, particularly in small mammals with hibernation. It has also been shown that fasted ruminants have significantly lower metabolic rates than when they are recently fed (Renecker and Hudson, 1985; Blaxter, 1989) so that feed restriction has to be considered as an effect on the ET. Also juvenile and adult animals might differ in their ET`s due to the higher FMR in growing animals.

Woodford et al. (1984) noted that the water turn over in early lactating cows is higher than in late prepartum and late lactating cows (Woodford et al., 1984) and such variations in water turn over rates are also likely to affect the ET.

Conclusion

Since the animal classes diverge in the ET for most isotopes the transfer of ETs between animal classes is highly limited. From our review it seems that there are too many other factors apart from BW influencing the time until equilibration of isotopes in the body compartments occurs. Consequently it appears justified to transfer ET results only if the species as well as the conditions are the same. The ETs in table 1 can serve as an orientation for future studies and the cohort and season mentioned should serve as additional information and should be considered in cases of ET transfer. Additional factors apart from BW and animal class should be considered in future studies, as they may influence the ET. With changed environment and animal cohorts the ET needs to be adapted. Hence authors should describe the animal and environmental data in more detailed.

References

Acquarone, M. and E. W. Born (2007). Estimation of water pool size, turnover rate and body composition of free-ranging Atlantic walruses (*Odobenus rosmarus rosmarus*) studied by isotope dilution. *J. Mar. Biol. Assoc. U.K.* 87, 77-84.

Adams, N. J., R. W. Abrams, W. R. Siegfried, K. A. Nagy and I. R. Kaplan (1991). Energy-expenditure and food-consumption by breeding cape gannets *Morus-capensis. Mar. Ecol.-Prog. Ser.* 70, 1-9.

Ambrose, S. J., S. D. Bradshaw, P. C. Withers and D. P. Murphy (1996). Water and energy balance of captive and free-ranging spinifexbirds (*Eremiornis carteri*) North (*Aves: Sylviidae*) on Barrow Island, Western Australia. *Aust. J. Zool.* 44, 107-117.

Anava, A., M. Kam, A. Shkolnik and A. A. Degen (2002). Seasonal daily, daytime and night-time field metabolic rates in Arabian babblers (*Turdoides squamiceps*). *J. Exp. Biol.* 205, 3571-3575.

Anbar, M. and Z. Lewitus (1958). Rate of body-water distribution studied with triple labelled water. *Nature* 181, 344-344.

Anderson, N. L., T. E. Hetherington and J. B. Williams (2003). Validation of the doubly labeled water method under low and high humidity to estimate metabolic rate and water flux in a tropical snake (*Boiga irregularis*). *J. Appl. Physiol.* 95, 184-191.

Andrews, F. M., J. A. Nadeau, L. Saabye and A. M. Saxton (1997). Measurement of total body water content in horses, using deuterium oxide dilution. *Am. J. Vet. Res.* 58, 1060-1064.

Blanc, S., A. Geloen, C. Pachiaudi, C. Gharib and S. Normand (2000). Validation of the doubly labeled water method in rats during isolation and simulated weightlessness. *Am. J. Physiol.-Reg. I.* 279, R1964-R1979.

Blaxter, K. L. (1989). Energy Metabolism in Animals and Man. Cambridge University Press, Cambridge, UK.

Born, H. J. (1945). Neuere biologische Anwendungen radioaktiver Isotope. *Angew. Chem.* 58, 45-45.

Bradshaw, S. D. and F. J. Bradshaw (2007). Isotopic measurements of field metabolic rate (FMR) in the marsupial honey possum (*Tarsipes rostratus*). *J. Mammal.* 88, 401-407.

Bradshaw, S. D., K. D. Morris, C. R. Dickman, P. C. Withers and D. Murphy (1994). Field metabolism and turnover in the golden bandicoot (*Isoodon-auratus*) and other small mammals from Barrow Island, Western-Australia. *Aust. J. Zool.* 42, 29-41.

Bryce, J. M., J. R. Speakman, P. J. Johnson and D. W. Macdonald (2001). Competition between Eurasian red and introduced Eastern grey squirrels: the energetic significance of body-mass differences. *P. Roy. Soc. Lond. B - Bio.* 268, 1731-1736.

Butte, N. F., W. W. Wong, P. D. Klein and C. Garza (1991). Measurement of milk intake - tracer-to-infant deuterium dilution method. *Brit. J. Nutr.* 65, 3-14.

Caire, G., A. M. Calderon de la Barca, A. V. Bolanos, M. E. Valencia, A. W. Coward, G. Salazar and E. Casanueva (2002). Measurement of deuterium oxide by infrared spectroscopy and isotope ratio mass spectrometry for quantifying daily milk intake in breastfed infants and maternal body fat. *Food Nutr Bull* 23, 38-41.

Chappell, M. A., V. H. Shoemaker, D. N. Janes, S. K. Maloney and T. L. Bucher (1993). Energetics of foraging in breeding Adelie penguins. *Ecology* 74, 2450-2461.

Christian, K., G. Bedford, B. Green, A. Griffiths, K. Newgrain and T. Schultz (1999). Physiological ecology of a tropical dragon, *Lophognathus temporalis*. *Aust. J. Ecol.* 24, 171-181.

Christian, K. A., G. Bedford, B. Green, T. Schultz and K. Newgrain (1998). Energetics and water flux of the marbled velvet gecko (*Oedura marmorata*) in tropical and temperate habitats. *Oecologia* 116, 336-342.

Christian, K. A., J. K. Webb and T. J. Schultz (2003). Energetics of bluetongue lizards (*Tiliqua scincoides*) in a seasonal tropical environment. *Oecologia* 136, 515-523.

Cole, T. J., W. A. Coward, M. Elia, C. J. Fjeld, M. Franklin, M. I. Goran, P. Haggarty, K. A. Nagy, A. M. Prentice, S. B. Roberts, D. A. Schoeller, K. Westerterp and W. W. Wong. (1990). The doubly-labelled water method for measuring energy expenditure: A consensus report by the IDECG working group. http://archive.unu.edu/unupress/food2/UID05E/UID05E00.HTM

Coleman, T. G., R. D. Manning, R. A. Norman and A. C. Guyton (1972). Dynamics of water-isotope distribution. *Am. J. Physiol.* 223, 1371-1375.

Corp, N., M. L. Gorman and J. R. Speakman (1999). Daily energy expenditure of free-living male Wood mice in different habitats and seasons. *Func. Ecol.* 13, 585-593.

Costa, D. P., J. P. Croxall and C. D. Duck (1989). Foraging energetics of Antarctic fur seals in relation to changes in prey availability. *Ecology* 70, 596-606.

Costa, D. P. and N. J. Gales (2000). Foraging energetics and diving behavior of lactating New Zealand sea lions, *Phocarctos hookeri. J. Exp. Biol.* 203, 3655-3665.

Costa, D. P. and N. J. Gales (2003). Energetics of a benthic diver: Seasonal foraging ecology of the Australian sea lion, *Neophoca cinerea. Ecol. Monogr.* 73, 27-43.

Culebras, J. M., G. F. Fitzpatrick, M. F. Brennan, C. M. Boyden and F. D. Moore (1977). Total-body water and exchangeable hydrogen 2. Review of comparative data from animals based on isotope dilution and desiccation, with a report of new data from rat. *Am. J. Physiol.* 232, R60-R65.

Degen, A. A., M. Kam, A. Hazan and K. A. Nagy (1986). Energy-expenditure and water flux in 3 sympatric desert rodents. *J. Anim. Ecol.* 55, 421-429.

Degen, A. A., B. Pinshow, P. U. Alkon and H. Arnon (1981). Tritiated water for estimating total body water and water turnover rate in birds. *J Appl Physiol* 51, 1183-1188.

Degen, A. A., B. Pinshow and M. Kam (1992). Field metabolic rates and water influxes of 2 sympatric gerbillidae - *Gerbillus-allenbyi* and *G-pyramidum. Oecologia* 90, 586-590.

Denny, M. J. S. and T. J. Dawson (1975). Comparative metabolism of tritiated-water by macropodid marsupials. *Am. J. Physiol.* 228, 1794-1799.

Dove, H. (1988): Estimation of the intake of milk by lambs, from the turnover of deuterium-labelled or tritium-labeled water. *Brit. J. Nutr.* 60, 375-387.

Dove, H. and M. Freer (1979). Accuracy of tritiated-water turnover rate as an estimate of milk intake in lambs. *Aust. J. Agr. Res.* 30, 725-739.

Drent, J., W. D. V. Lichtenbelt and M. Wikelski (1999). Effects of foraging mode and season on the energetics of the marine iguana, *Amblyrhynchus cristatus. Funct. Ecol.* 13, 493-499.

Dryden, G., B. Green, D. King and J. Losos (1990). Water and energy turnover in a small monitor lizard, *Varanus-acanthurus. Aust. Wildlife Res.* 17, 641-646.

Dykstra, C. R. and W. H. Karasov (1993). Daily energy-expenditure by nestling house wrens. *Condor* 95, 1028-1030.

Dziewiatkowski, D. D. (1951). Radioautographic visualization of sulfur-35 disposition in the articular cartilage and one of suckling rats following injection of labeled sodium sulfate. *J. Exp. Med.* 93, 451-458.

Eichhorn, G. and G. H. Visser (2008). Evaluation of the deuterium dilution method to estimate body composition in the barnacle goose: accuracy and minimum equilibration time. Physiol. *Biochem. Zool.* 81, 508-518.

Ellis, W. A. H., A. Melzer, B. Green, K. Newgrain, M. A. Hindell and F. N. Carrick (1995). Seasonal-variation in water flux, field metabolic-rate and food-consumption of free-ranging Koalas (*Phascolarctos-cinereus*). *Aust. J. Zool.* 43, 59-68.

Engel, S., H. Biebach and G. H. Visser (2006). Metabolic costs of avian flight in relation to flight velocity: a study in rose coloured starlings (*Sturnus roseus, Linnaeus*). *J. Comp. Physiol. B* 176, 415-427.

Evans, M., B. Green and K. Newgrain (2003). The field energetics and water fluxes of free-living wombats (*Marsupialia : Vombatidae*). *Oecologia* 137, 171-180.

Fancy, S. G., J. M. Blanchard, D. F. Holleman, K. J. Kokjer and R. G. White (1986). Validation of doubly labeled water method using a ruminant. *Am. J. Physiol.* 251, R143-R149.

Farrell, D. J., J. L. Corbett and R. A. Leng (1972). Undernutrition in grazing sheep 2. Calorimetric measurements on sheep taken from pasture. *Aust. J. Agr. Res.* 23, 499-509.

Fleischmann, R. (1945). Allgemeine Anwendung der stabilen und radioaktiven Isotope. *Angew. Chem.* 58: 45-45.

Fowler, M. E. (2010). Medicine and Surgery of Camelides. 3rd Edition. Wiley-Blackwell, Oxford, UK.

Fowler, S. L., D. P. Costa, J. P. Y. Arnould, N. J. Gales and J. M. Burns (2007). Ontogeny of oxygen stores and physiological diving capability in Australian Sea Lions. *Funct. Ecol.* 21, 922-935.

Fuller, Z., C. A. Maltin, E. Milne, G. S. Mollison, J. E. Cox and C. M. Argo (2004). Comparison of calorimetry and the doubly labelled water technique for the measurement of energy expenditure in *Equidae. Anim. Sci.* 78, 293-303.

Fyhn, M., G. W. Gabrielsen, E. S. Nordoy, B. Moe, I. Langseth and C. Bech (2001). Individual variation in field metabolic rate of kittiwakes (*Rissa tridactyla*) during the chick-rearing period. *Physiol. Biochem. Zool.* 74, 343-355.

Gales, R. and B. Green (1990). The annual energetics cycle of little penguins (*Eudyptula-minor*). *Ecology* 71, 2297-2312.

Garcia, C., L. Rosetta, A. Ancel, P. C. Lee and M. Caloin (2004). Kinetics of stable isotope and body composition in olive baboons (*Papio anubis*) estimated by deuterium dilution space: a pilot study. *J. Med. Primatol.* 33, 146-151.

Geffen, E., A. A. Degen, M. Kam, R. Hefner and K. A. Nagy (1992). Daily energy-expenditure and water flux of free-living Blanford's foxes (*Vulpes-cana*), a small desert carnivore. *J. Anim. Ecol.* 61, 611-617.

Gessaman, J. A., K. Newgrain and B. Green (2004). Validation of the doubly-labeled water (DLW) method for estimating CO_2 production and water flux in growing poultry chicks. *J. Avian Biol.* 35, 71-96.

Gilbert, J. H., P. A. Zollner, A. K. Green, J. L. Wright and W. H. Karasov (2009). Seasonal field metabolic rates of American martens in Wisconsin. *Am. Midl. Nat.* 162, 327-334.

Godfrey, J. D., D. M. Bryant and M. J. Williams (2003). Radio-telemetry increases free-living energy costs in the endangered Takahe *Porphyrio mantelli*. *Biol. Conserv.* 114, 35-38.

Gorman, M. L., M. G. Mills, J. P. Raath and J. R. Speakman (1998). High hunting costs make African wild dogs vulnerable to kleptoparasitism by hyaenas. *Nature* 391, 479-481.

Gotaas, G., E. Milne, P. Haggarty and N. J. Tyler (1997). Use of feces to estimate isotopic abundance in doubly labeled water studies in reindeer in summer and winter. *Am J Physiol* 273, R1451-R1456.

Gotaas, G., E. Milne, P. Haggarty and N. J. Tyler (2000). Energy expenditure of free-living reindeer estimated by the doubly labelled water method. *Rangifer* 2, 211-219.

Green, B., G. Dryden and K. Dryden (1991a). Field energetics of a large carnivorous lizard, *Varanus-rosenbergi*. *Oecologia* 88, 547-551.

Green, B., D. King and A. Bradley (1989). Water and energy-metabolism and estimated food-consumption rates of free-living wambengers, *Phascogale-calura (Marsupialia, Dasyuridae)*. *Aust. Wildlife Res.* 16, 501-507.

Green, B., D. King, M. Braysher and A. Saim (1991b). Thermoregulation, water turnover and energetics of free-living komodo dragons, *Varanus-komodoensis*. *Comp. Biochem. Physiol. A* 99, 97-101.

Green, B., D. King and H. Butler (1986). Water, sodium and energy turnover in free-living perenties, *Varanus-giganteus*. *Aust. Wildlife Res.* 13, 589-595.

Haggarty, P., J. J. Robinson, J. Ashton, E. Milne, C. L. Adam, C. E. Kyle, S. L. Christie and A. J. Midwood (1998). Estimation of energy expenditure in free-living red deer (*Cervus elaphus*) with the doubly-labelled water method. *Brit. J. Nutr.* 80, 263-272.

Hawkins, P. A. J., P. J. Butler, A. J. Woakes and J. R. Speakman (2000). Estimation of the rate of oxygen consumption of the common eider duck (*Somateria mollissima*), with some measurements of heart rate during voluntary dives. *J. Exp. Biol.* 203, 2819-2832.

Hinchcliff, K. W., G. A. Reinhart, J. R. Burr, C. J. Schreier and R. A. Swenson (1997). Metabolizable energy intake and sustained energy expenditure of Alaskan sled dogs during heavy exertion in the cold. *Am. J. Vet. Res.* 58, 1457-1462.

Humphries, M. M., S. Boutin, D. W. Thomas, J. D. Ryan, C. Selman, A. G. McAdam, D. Berteaux and J. R. Speakman (2005). Expenditure freeze: the metabolic response of small mammals to cold environments. *Ecol. Lett.* 8, 1326-1333.

Jones, T. T., M. D. Hastings, B. L. Bostrom, R. D. Andrews and D. R. Jones (2009). Validation of the use of doubly labeled water for estimating metabolic rate in the green turtle (*Chelonia mydas L.*): a word of caution. *J. Exp. Biol.* 212, 2635-2644.

Karasov, W. H. (1981). Daily energy-expenditure and the cost of activity in a free-living mammal. *Oecologia* 51, 253-259.

Kenagy, G. J., S. M. Sharbaugh and K. A. Nagy (1989). Annual cycle of energy and time expenditure in a golden-mantled ground-squirrel population. *Oecologia* 78, 269-282.

Kooyman, G. L., Y. Cherel, Y. Lemaho, J. P. Croxall, P. H. Thorson and V. Ridoux (1992). Diving behavior and energetics during foraging cycles in king penguins. *Ecol. Monogr.* 62, 143-163.

Krol, E. and J. R. Speakman (1999). Isotope dilution spaces of mice injected simultaneously with deuterium, tritium and oxygen-18. *J. Exp. Biol.* 202, 2839-2849.

Lifson, N., G. B. Gordon and R. McClintock (1955). Measurement of total carbon dioxide production by means of $D_2^{18}O$. *J. Appl. Physiol.* 7, 704-710.

Lifson, N. and J. S. Lee (1959). Measurement of energy and material balance by the use of doubly-labeled water. *Fed. Proc.* 18, 93-93.

Loeffler, K. (2002). Anatomie und Physiologie der Haustiere. 10th Edition. Verlag Eugen Ulmer, Stuttgart, Germany

Lynn, S. E., A. M. Houtman, W. W. Weathers, E. D. Ketterson and V. Nolan (2000). Testosterone increases activity but not daily energy expenditure in captive male dark-eyed juncos, *Junco hyemalis. Anim. Behav.* 60, 581-587.

MacFarlane, W. V., B. Howard and B. D. Siebert (1969). Tritiated water in measurement of milk intake and tissue growth of ruminants in field. *Nature* 221, 578-579.

Martin, L., B. Siliart, H. Dumon, R. Backus, V. Biourge and P. Nguyen (2001). Leptin, body fat content and energy expenditure in intact and gonadectomized adult cats: a preliminary study. *J. Anim. Physiol. Anim Nutr* 85, 195-199.

Mattauch, J. (1947). Stabile Isotope - Ihre Messung und Ihre Verwendung. *Angew. Chem.* 59, 37-42.

McClintock, R. and N. Lifson (1958a). CO_2 output of mice measured by $D_2^{18}O$ under conditions of isotope re-entry into the body. *Am. J. Physiol.* 195, 721-725.

McClintock, R. and N. Lifson (1958b). Determination of the total carbon dioxide outputs of rats by the $D_2^{18}O$ method. *Am. J. Physiol.* 192, 76-78.

McClintock, R. and N. Lifson (1958c). Measurement of basal and total metabolism in hereditarily obese-hyperglycemic mice. *Am. J. Physiol.* 193, 495-498.

Mehlum, F., G. W. Gabrielsen and K. A. Nagy (1993). Energy-expenditure by Black Guillemots (*Cepphus-grylle*) during chick-rearing. Colon. *Waterbird.* 16, 45-52.

Midwood, A. J., P. Haggarty and B. A. McGaw (1993). The doubly labeled water method - errors due to deuterium-exchange and sequestration in ruminants. *Am. J. Physiol.* 264, R561-R567.

Midwood, A. J., P. Haggarty, B. A. McGaw, G. S. Mollison, E. Milne and G. J. Duncan (1994). Validation in sheep of the doubly labeled water method for estimating Co_2 production. *Am. J. Physiol.* 266, R169-R179.

Midwood, A. J., P. Haggarty, B. A. McGaw and J. J. Robinson (1989). Methane production in ruminants - Its effect on the doubly labeled water method. *Am. J. Physiol.* 257, R1488-R1495.

Munks, S. A. and B. Green (1995). Energy allocation for reproduction in a marsupial arboreal folivore, the common ringtail possum (*Pseudocheirus-peregrinus*). *Oecologia* 101, 94-104.

Mutze, G. J., B. Green and K. Newgrain (1991). Water flux and energy use in wild house mice (*Mus-Domesticus*) and the impact of seasonal aridity on breeding and population-levels. *Oecologia* 88, 529-538.

Nagy, K. A. and S. D. Bradshaw (2000). Scaling of energy and water fluxes in free-living arid-zone Australian marsupials. *J. Mammal.* 81, 962-970.

Nagy, K. A., S. D. Bradshaw and B. T. Clay (1991). Field metabolic-rate, water flux, and food-requirements of short-nosed bandicoots, Isoodon-Obesulus (*Marsupialia, Peramelidae*). *Aust. J. Zool.* 39, 299-305.

Nagy, K. A. and D. P. Costa (1980). Water flux in animals - analysis of potential errors in the tritiated-water method. *Am. J. Physiol.* 238, R454-R465.

Nagy, K. A., A. K. Lee, R. W. Martin and M. R. Fleming (1988). Field metabolic-rate and food requirement of a small dasyurid marsupial, *Sminthopsis-crassicaudata*. *Aust. J. Zool.* 36, 293-299.

Nagy, K. A. and R. W. Martin (1985). Field metabolic-rate, water flux, food-consumption and time budget of koalas, *Phascolarctos-cinereus (Marsupialia, Phascolarctidae)* in Victoria. *Aust. J. Zool.* 33, 655-665.

Nagy, K. A. and K. Milton (1979). Energy-metabolism and food-consumption by wild Howler monkeys (*Alouatta-palliata*). *Ecology* 60, 475-480.

Nagy, K. A., G. D. Sanson and N. K. Jacobsen (1990). Comparative field energetics of 2 macropod marsupials and a ruminant. *Aust. Wildlife Res.* 17, 591-599.

Obst, B. S. and K. A. Nagy (1992). Field energy expenditures of the Southern giant-petrel. *Condor* 94, 801-810.

Peacock, W. L., E. Krol, K. M. Moar, J. S. McLaren, J. G. Mercer and J. R. Speakman (2004). Photoperiodic effects on body mass, energy balance and hypothalamic gene expression in the bank vole. *J. Exp. Biol.* 207, 165-177.

Quin, D. G., A. Riek, S. Green, A. P. Smith and F. Geiser (2010). Seasonally constant field metabolic rates in free-ranging sugar gliders (*Petaurus breviceps*). *Comp. Biochem. Physiol. A* 155, 336-340.

Rabideau, G. S. and G. O. Burr (1945). The use of the C-13 isotope as a tracer for transport studies in plants. *Am. J. Bot.* 32, 349-356.

Renecker, L. A. and R. J. Hudson (1985). Telemetered heart-rate as an index of energy-expenditure in moose (*Alces-alces*). *Comp. Biochem. Physiol. A* 82, 161-165.

Riek, A. (2006). Investigations on milk composition; milk intake and body weight development in llama (*lama glama*). Doctoral dissertation, University of Goettingen, Cuvillier Verlag.

Riek, A. (2008). Relationship between field metabolic rate and body weight in mammals: effect of the study. *J. Zool.* 276, 187-194.

Riek, A., M. Gerken and E. Moors (2007a). Measurement of milk intake in suckling llamas (*Lama glama*) using deuterium oxide dilution. *J. Dairy Sci.* 90, 867-875.

Riek, A., L. Van Der Sluijs and M. Gerken (2007b). Measuring the energy expenditure and water flux in free-ranging alpacas (*Lama pacos*) in the Peruvian Andes using the doubly labelled water technique. *J. Exp. Zool.* 307A, 667-675.

Roe, J. H., A. Georges and B. Green (2008). Energy and water flux during terrestrial estivation and overland movement in a freshwater turtle. *Physiol. Biochem. Zool.* 81, 570-583.

SAS - Statistical Analysis System. (2009). User's guide 9.1.3. Cary: NC, SAS Institute Inc.

Salsbury, C. M. and K. B. Armitage (1994). Resting and field metabolic rates of adult male yellow-bellied marmots, *Marmota-flaviventris. Comp. Biochem. Physiol. A* 108, 579-588.

Scantlebury, M. and R. Butterwick (2001). Energetics and litter size variation in domestic dog *Canis familiaris* breeds of two sizes. *Comp. Biochem. Physiol. A* 129, 919-931.

Speakman, J.R. (2010): Doubly Labelled Water Resource Centre. University of Aberdeen. http://www.abdn.ac.uk/energetics-research/doubly-labelled-water

Speakman, J. R, M. Scantlebury, A. F. Russell, G. M. McIlrath, J. R. Speakman and T. H. Clutton-Brock (2002). The energetics of lactation in cooperatively breeding meerkats *Suricata suricatta. P. Roy. Soc. Lond. B - Bio.* 269, 2147-2153.

Scantlebury, M., J. M. Waterman, M. Hillegass, J. R. Speakman and N. C. Bennett (2007). Energetic costs of parasitism in the cape ground squirrel *Xerus inauris. P. Roy. Soc. B-Bio.* 274, 2169-2177.

Schmid, J. and J. R. Speakman (2000). Daily energy expenditure of the grey mouse lemur (*Microcebus murinus*): a small primate that uses torpor. *J. Comp. Physiol. B* 170, 633-641.

Schmidt-Nielsen, K., E. C. Crawford, A. E. Newsome, K. S. Rawson and H. T. Hammel (1967). Metabolic rate of camels - Effect of body temperature and dehydration. *Am. J. Physi.* 212, 341-346.

Schoeller, D. A. and E. Vansanten (1982). Measurement of energy-expenditure in humans by doubly labeled water method. *J. Appl. Physiol.* 53: 955-959.

Siebert, B. D. and W. V. MacFarlane (1971). Water turnover and renal function of dromedaries in the desert. *Physiol. Zool.* 44, 225-240.

Smith, A. P., K. A. Nagy, M. R. Fleming and B. Green (1982). Energy-requirements and water turnover in free-living leadbeater possums, *Gymnobelideus-leadbeateri* (*Marsupialia, Petauridae*). *Aust. J. Zool.* 30, 737-749.

Smith, B. S. W. and A. R. Sykes (1974). Effect of route of dosing and method of estimation of tritiated-water space on determination of total-body water and prediction of body fat in sheep. *J. Agr. Sci.* 82, 105-112.

Sparling, C. E., D. Thompson, M. A. Fedak, S. L. Gallon and J. R. Speakman (2008). Estimating field metabolic rates of pinnipeds: doubly labelled water gets the seal of approval. *Funct. Ecol.* 22, 245-254.

Speakman, J. (2010). Doubly Labelled Water - Resource Centre. http://www.abdn.ac.uk/energetics-research/doubly-labelled-water.20.08.2010

Speakman, J. R. and B. Krol (2005). Comparison of different approaches for the calculation of energy expenditure using doubly labeled water in a small mammal. *Physiol. Biochem. Zool.* 78, 650-667.

Speakman, J. R., G. Perez-Camargo, T. McCappin, T. Frankel, P. Thomson and V. Legrand-Defretin (2001). Validation of the doubly-labelled water technique in the domestic dog (*Canis familiaris*). *Brit. J. Nutr.* 85, 75-87.

Stephenson, P. J., J. R. Speakman and P. A. Racey (1994). Field metabolic-rate in 2 species of shrew-tenrec, *Microgale-Dobsoni* and *M-Talazaci*. *Comp. Biochem. Physiol. A* 107, 283-287.

Swindlehurst, E. and E. G. Woodroofe (1952). Application of the isotope tracer technique in the manufacture of animal feeding stuffs. *Med Biol Illus* 2, 51-60.

Theil, R. K., N. B. Kristensen, H. Jorgensen, R. Labouriau and K. Jakobsen (2007). Milk intake and carbon dioxide production of piglets determined with the doubly labelled water technique. *Animal* 1, 881-888.

Thompson, G. G., S. D. Bradshaw and P. C. Withers (1997). Energy and water turnover rates of a free-living and captive goanna, *Varanus caudolineatus* (*Lacertilia: Varanidae*). *Comp. Biochem. Physiol. A* 116, 105-111.

Toerien, C. A., T. Sahlu and W. W. Wong (1999). Energy expenditure of Angora bucks in peak breeding season estimated with the doubly-labeled water technique. *J Anim. Sci.* 77, 3096-3105.

Trillmich, F. and G. L. Kooyman (2001). Field metabolic rate of lactating female Galapagos fur seals (*Arctocephalus galapagoenis*): the influence of offspring age and environment. *Comp. Biochem. Physiol. A* 129, 741-749.

Tyndale-Biscoe, H. (2005). What is a marsupial? In Life of marsupials: 1–36. Tyndale-Biscoe, H. (Ed.). CSIRO Publishing, Collingwood, Australia:

van Marken Lichtenbelt, W. D., R. A. Wesselingh, J. T. Vogel and K. B. M. Albers (1993). Energy budgets in free-living green iguanas in a seasonal environment. *Ecology* 74, 1157-1172.

van Trigt, R., E. R. T. Kerstel, R. E. M. Neubert, H. A. J. Meijer, M. McLean and G. H. Visser (2002). Validation of the DLW method in Japanese quail at different water fluxes using laser and IRMS. *J. Appl. Physiol.* 93, 2147-2154.

Visser, G. H., A. Dekinga, B. Achterkamp and T. Piersma (2000). Ingested water equilibrates isotopically with the body water pool of a shorebird with unrivaled water fluxes. *Am. J. Physiol.-Reg. I.* 279, R1795-R1804

Wallis, I. R., B. Green and K. Newgrain (1997). Seasonal field energetics and water fluxes of the long-nosed potoroo (*Potorous tridactylus*) in Southern Victoria. *Aust. J. Zool.* 45,1-11.

Weathers, W. W., W. A. Buttemer, A. M. Hayworth and K. A. Nagy (1984). An evaluation of time-budget estimates of daily energy-expenditure in birds. *Auk* 101, 459-472.

Weathers, W. W. and K. A. Nagy (1984). Daily energy-expenditure and water flux in black-rumped waxbills (*Estrilda-troglodytes*). *Comp. Biochem. Phys. A* 77, 453-458.

Weathers, W. W., D. C. Paton and R. S. Seymour (1996). Field metabolic rate and water flux of nectarivorous honeyeaters. *Aust. J. Zool.* 44, 445-460.

Weiss, P., H. Wang, A. C. Taylor and M. V. Edds (1945). Proximo-distal fluid convection in the endoneurial spaces of peripheral nerves, demonstrated by colored and radioactive (Isotope) tracers. *Am. J. Physiol.* 143, 521-540.

Welcker, J., B. Moe, C. Bech, M. Fyhn, J. Schultner, J. R. Speakman and G. W. Gabrielsen (2010). Evidence for an intrinsic energetic ceiling in free-ranging kittiwakes *Rissa tridactyla. J. Anim. Ecol.* 79, 205-213

Williams, B. (2001). Seasonal variation in energy expenditure, water flux and food consumption of Arabian oryx *oryx leucoryx. J. Exp. Biol.* 204, 2301-2311.

Williams, J. B., M. D. Anderson and P. R. K. Richardson (1997). Seasonal differences in field metabolism, water requirements, and foraging behavior of free-living aardwolves. *Ecology* 78, 2588-2602.

Williams, J. B., W. R. Siegfreid, S. J. Milton, N. J. Adams, W. R. J. Dean, M. A. Duplessis, S. Jackson and K. A. Nagy (1993). Field metabolism, water requirements, and foraging behavior of wild ostriches in the Namib. *Ecology* 74, 390-404.

Winstanley, R. K., W. A. Buttemer and G. Saunders (2003). Field metabolic rate and body water turnover of the red fox *Vulpes vulpes* in Australia. *Mammal Rev.* 33, 295-301.

Woodford, S. T., M. R. Murphy and C. L. Davis (1984). Water dynamics of dairy-cattle as affected by initiation of lactation and feed-intake. *J. Dairy Sci.* 67, 2336-2343.

Yagil, R. and Z. Etzion (1980). Milk-yield of camels (*Camelus-Dromedarius*) in drought areas. *Comp. Biochem. Physiol.* A 67, 207-209.

Znari, M. and K. A. Nagy (1997). Field metabolic rate and water flux in free-living Bibron's agama (*Agama impalearis, Boettger,* 1874) in Morocco. *Herpetologica* 53, 81-88.

CHAPTER 5

Seasonal changes in total body water and water intake of Shetland ponies measured by an isotope dilution technique

L. Brinkmann, M. Gerken, A. Riek

Department of Animals Sciences, University of Goettingen, Albrecht-Thaer Weg 3, 37075 Goettingen, Germany

Seasonal changes in total body water and water intake of Shetland ponies measured by an isotope dilution technique

Lea Brinkmann, Martina Gerken, Alexander Riek

Department of Animals Sciences, University of Goettingen,
Albrecht-Thaer Weg 3, 37075 Goettingen, Germany

Abstract

Water can be considered as one of the most essential nutrients necessary to support life and adequate water supply is crucial for animal survival and productivity. The present study was designed to determine seasonal changes in the water metabolism of horses under extensive housing conditions. Total body water (TBW) and total water intake (TWI) of 10 adult Shetland pony mares were estimated at monthly intervals for 14 month by using the Deuterium dilution technique. During the last 4 month, 5 animals were fed restrictively to simulate natural feed shortage in winter, while 5 animals served as control group. The TBW (kg) was closely related to body mass (TBW, kg = -2.86 + 0.67·body mass, kg, $P < 0.001$, $N = 105$) explaining 86% of the variation. In contrast, TBW (%) remained relatively stable across all measurements (range: 57.8 - 71.2%). The TWI was higher in summer than in winter with a mean of 87.4 ± 28.9 $mL·kg^{-1}·d^{-1}$. However, TWI measured at ambient temperatures (T_a) < 0 °C did not follow the same trend as TWI at T_a's > 0 °C. Therefore, removing TWI values measured at T_a's < 0 °C from the analysis resulted in highly significant correlations with locomotor activity ($r = 0.87$), T_a ($r = 0.86$) and resting heart rate ($r = 0.88$). However, only heart rate and T_a had a significant effect on TWI. The multiple regression between TWI, T_a and heart rate explained 84% of the variation in TWI (TWI, $mL·kg^{-1}·d^{-1}$ = -13.38 + 1.77·heart rate, beats·min^{-1} + 2.11·T_a, °C, $P < 0.001$). Feed restriction had no effect on TWI and TBW. The TBW content was unaffected by season and physical activity. The established regression equation for TBW and body mass can be used to predict TBW from body mass in ponies under field conditions. The comparison of TWI with published data on drinking water intake revealed that ponies had 1.7 to 5.1 times higher total water intakes, when other sources of water such as feed and metabolic water were included. The TWI was highly influenced by environmental conditions and metabolic rate. Contrary to expectation, water supply during the cold seasons might be more critical than under summer conditions, when water content of grass is high to allow compensating for limited availability of drinking water.

Keywords: body water, deuterium oxide, feed restriction, horses, seasonal changes, water turnover

Introduction

Water can be considered as one of the most essential nutrients necessary to support life. The water demand of an animal is influenced by various factors, such as species, body mass, breed, feeding regime, physical activity etc. (Crowell-Davis et al., 1985; Cymbaluk, 1990; Kristula and McDonnell, 1994; Scheibe et al., 1998; Hoffmann et al., 2009) and may change considerably under extensive outdoor housing conditions because food composition, ambient temperature (T_a) and solar radiation vary remarkably with seasons.

The direct measurement of drinking water does not account for other sources of water intake such as preformed water in feed and metabolic water and thus will underestimate the total water flux. In this context, the isotope dilution technique (Lifson and McClintock, 1966) is a suitable method for estimating individual water flux in free ranging horses when kept under extensive housing conditions. The total body water (TBW) content is an important parameter when determining drug doses and fluid therapies in case of dehydration (Fielding et al., 2004; Riek and Gerken, 2010). The TBW is also used for assessing body composition in live animals (Robb et al., 1972) and can be influenced by e.g., dehydration and body condition (Bakker and Main, 1980; Sneddon et al., 1991) or long-term feed restriction as feed shortage leads to changes in body composition.

A few studies measured the TBW content of horses (e.g., Andrews et al., 1997; Forro et al., 2000), but to our knowledge long term studies on total water intake (TWI) and TBW in horses are lacking. Therefore, the aim of our study was to measure TWI continuously over an entire year using the isotope dilution technique. We were especially interested in the modulating effects of climate, locomotor behavior, heart rate and food supply and how they affect TWI.

Materials, methods and animals

The research conducted in this study was performed in accordance with the current laws regulating animal welfare and experiments with animals in Germany and was approved by the State Agency of Consumer Protection and Food Safety of Lower Saxony.

Animals and Management

The study was conducted at the Department of Animal Sciences at the University of Göttingen, Germany, lasting 14 month from January 2010 until February 2011. Initially, we started with 8 Shetland pony mares, however, the herd was increased to 10 mares at the beginning of May 2010. The age of the animals ranged from 4 to 16 years and body mass over the entire experimental period was 148.5 ± 35.6 kg (mean ± SD). Animals originated from different farms in Germany and the Netherlands and were used to outdoor housing systems.

From January 2010 until the end of May 2010 ponies were housed in 2 groups of 5 ponies each on two 210 m^2 paddocks. Each paddock had a permanent access to a 18.9 m^2 pen that was covered with straw. Two large exits allowed animals to enter and leave without rank conflicts. Both identical pens were equipped with 5 feeding stands each (1.35 m x 1.60 m x 0.55 m; height x length x width) ensuring that each horse was able to ingest the required food. The pens were not heated and hence the temperature inside the stable was comparable with the ambient temperature. The dark-light cycle inside the stable fluctuated according to the natural photoperiod. From January 2010 until the end of May 2010 all horses were subjected to the same feeding treatment receiving straw *ad libitum* and daily 5 kg hay per 100 kg body mass. Additionally ponies received daily 22 g mineral mixture per 100 kg body mass

(Derby® Mineralpellets, Derby Spezialfutter GmbH, Münster, Germany) and occasionally small amounts of concentrate (Derby® Standard, Derby Spezialfutter GmbH, Münster, Germany). Water was available *ad libitum* throughout the experiment at frost prove drinking troughs.

From the end of May 2010 until the middle of Oct 2010 all animals were kept on a permanent pasture with access to two 4.0 m x 3.6 m shelters. In addition to pasture, ponies were offered small amounts of hay, straw, and mineral feed. In the middle of October 2010 ponies were brought back into the stable where they were housed in the same pens described above until the end of our study (February 2011). During that time ponies were allocated according to their body mass and BCS (Carroll and Huntington, 1988) into a control and a treatment group of 5 animals each, resulting in comparable mean body mass and BCS for both groups. In the first 2 weeks both groups were fed identically as described above. However, while from the beginning of November 2010 the control group received the same diet as before, the experimental group was fed restrictively to simulate limited availability of feed in autumn and winter under natural conditions. The amount of feed offered in the restricted animals was gradually reduced from 100% to 80% of the recommended energy and protein requirements for Shetland ponies (GfE, 1994) until 21 January and to a further 70% until 28 February 2011.

Over the entire study period body mass was recorded before and after each isotope measurement period (see below) while BSC was evaluated on a monthly basis.

Total Water Intake

The TBW and TWI of ponies were determined at monthly intervals during the entire experimental period using the deuterium oxide (D_2O) dilution technique (Lifson and McClintock, 1966). The dilution space of the isotope is considered to reflect TBW content which is determined by the dilution and equilibration of the labeled water with body fluids. The water turnover rate is calculated from the wash out of the isotope over a known period of time. Prior to the D_2O administration a blood sample of 5 mL was drawn from the *vena jugularis* from every animal to specify the natural abundance of ^2H in body fluids. Subsequently the animals were injected intravenously with D_2O of 99.90 % purity (0.1 $g \cdot kg^{-1}$; Euriso-top GmbH, Saarbrücken, Germany). The individual amount administrated was determined by the mass of the animal prior to the application, which was measured with a scale (Sartorius CW3P1-

150IG-1, Sartorius AG, Göttingen, Germany). The actual dose given was gravimetrically measured by weighing the syringe before and after the administration to the nearest 0.001 g (Sartorius model CW3P1-150IG-1, Sartorius AG, Göttingen, Germany). Fifty-four and 72 hours after dosing, blood samples were taken from the *vena jugularis* into blood tubes containing sodium citrate. Blood samples were centrifuged at 3500 rpm for 10 minutes (centrifugation force: 1620 g) and the resulting plasma was pipetted into 0.7 mL glass vials and stored at $-20\,°C$ until determination of D_2O concentration. Analyses for ^2H were carried out at the Competence Centre for Stable Isotopes (KOSI), University of Göttingen, Germany. Isotope ratios of ^2H were measured using an on-line high temperature reduction technique in a helium carrier gas described previously (Gehre et al., 2004) and expressed relative to the Vienna standard mean ocean water (VSMOW), which is the international reference standard for D_2O. The averages of 5 measurements per sample were calculated. The isotopic equilibrium concentration and fractional water turnover were calculated individually for all animals and measurements. Isotopic equilibrium concentration and fractional water turnover were computed for each animal by extrapolating the regression of D_2O concentrations (C_t) on time by the equation:

$$C_t = C_0 \cdot e^{-k \cdot t}$$
(Eq. 1)

where C_0 = equilibration concentration (intercept) for D_2O, k = fractional water turn over (slope) and t = time elapsed since tracer administration (Hollemann et al., 1982). Isotope equilibration and pre-dose baseline concentration were then used to calculate the D_2O dilution space (N) in kg by using the formula from Schoeller et al. (1986):

$$N = [D \cdot APE_{dose} \cdot 18.02 \text{ g mole}^{-1}] : [MW_{dose} \cdot 100 \cdot (C_0 - C_b) \cdot R_{std}]$$
(Eq. 2)

where D = dose given in grams, APE_{dose} = atomic enrichment of the dose in per cent (99.90%), MW_{dose} = molecular weight of the dose (D_2O = 20.02), C_0 = equilibration concentration, C_b = pre dose baseline concentration, R_{std} = ratio of Deuterium and

Hydrogen of SMOW. Finally, the total water intake (TWI) was calculated after Oftedal et al. (1983):

$$TWI = L + G = k \cdot N$$
(Eq. 3)

where TWI = daily total water intake, L = amount of daily water lost, G = amount of daily water stored, k = daily water turnover rate. TWI thus includes drinking water, preformed water ingested by food and metabolic water.

Heart Rate

Heart rate was recorded every 2 weeks with a stethoscope between 1200 and 1300 h three times for 60 s. Before measurements animals were at rest for at least 5 minutes while loosely tied with a rope but with visual contact to their conspecifics. The ponies were used to the handling and thus any impact of the measuring procedure on heart rate recordings was assumed to be minimal.

Locomotor Activity and Ambient Temperature

To record the locomotor activity (LA), ALT-pedometers (Activity-Lying-Temperature-Pedometer; Engineering Office Holz, Falkenhagen, Germany) were fitted to ponies. Pedometers included 4 sensors that allowed the recording of LA and lying time. The measured LA consisted of activity impulses generated by the front leg and was registered continuously, with a maximum of 2 impulses recorded per s. The data were summarized and stored in 15 minutes intervals. Pedometers were lined with a silicon pad to avoid pressure marks on the foreleg. The T_a and humidity were recorded continuously every 10 minutes by 2 data loggers (Tiny view TV 1500, temperature resolution: 0.25 °C, humidity resolution: 0.5%, Gemini, Chichester, West Sussex, UK) over the entire experimental period.

Statistical analyses

Statistical analyses were performed using the software package Statistical Analyses System version 9.2 (SAS, 2008). A non-linear regression procedure was used to extrapolate the regression of C_t on time. A total of 110 calculated TBW and TWI measurements were included in the analysis. Heart rate, LA and body mass were arranged into classes resulting in 15 classes for heart rate and 14 classes for LA. The TBW and TWI data were subjected to an ANOVA using a non-linear variance model with body mass as a covariate. The model was:

$$Y_{ijklmn} = \mu + H_i + T_j + L_k + G_l + A_m + b_{ijklm}(BM) + e_{ijklmn}$$

where Y_{ijklmn}: observation value; μ: overall mean; H_i: fixed effect of the heart rate class; T_j: fixed effect of the T_a class; L_k: fixed effect of the LA class; G_l: fixed effect of the treatment group; A_m = random effect of the animal; $b_{ijklm}(BM)$: linear regression coefficient on body mass and e_{ijklm}: random error.

Results

Body mass varied over the course of the year for all ponies with the lowest values recorded in February (138 ± 36 kg), and the highest in October (158 ± 38 kg). The BCS followed a similar pattern with peak values in October (4.1 ± 0.7 points) and lower scores in February (3.1 ± 1.0 points). Restrictively fed animals decreased their body mass and BSC significantly in winter monthwhen feed was gradually reduced (BCS, points = 5.4 - 0.73·mo, R^2 = 0.65, $P < 0.001$; body mass, kg = 165.7 - 4.1·mo, R^2 = 0.12, $P < 0.05$), whereas in the control mares body mass and BSC did not change (BCS, points = 3.4 - 0.09·mo, R^2 = 0.03, $P = 0.43$; body mass, kg = 149.8 - 0.22·mo, R^2 = 0.05, $P = 0.93$).

There were no significant differences in LA, TBW (%), TBW (kg) and mass-specific TWI (i.e. mL·kg^{-1}·d^{-1}) between the control and feed restricted animals over the entire feeding trial. The TBW (%) over the entire study period ranged from 50.9 to 78.3. TBW (%) values in June 2010 were significantly higher (71.2 ± 5.0%) than those in September 2010 (57.8 ± 1.8%), January 2010 (62.3 ± 6.3%) or February 2011 (58.2 ± 5.2%). Heart rate, T_a and LA had no significant effect on TBW (%). The relationship between TBW (kg) and body mass was highly significant (TBW, kg = -2.86 + 0.67·body mass, kg, $P < 0.001$), explaining 86% of the variation in TBW (kg) (Figure 1). In contrast, TBW (%) was not affected by body mass (TBW, % = 43.8 + 0.14·body mass, kg, $P > 0.05$, R^2 = 0.06).

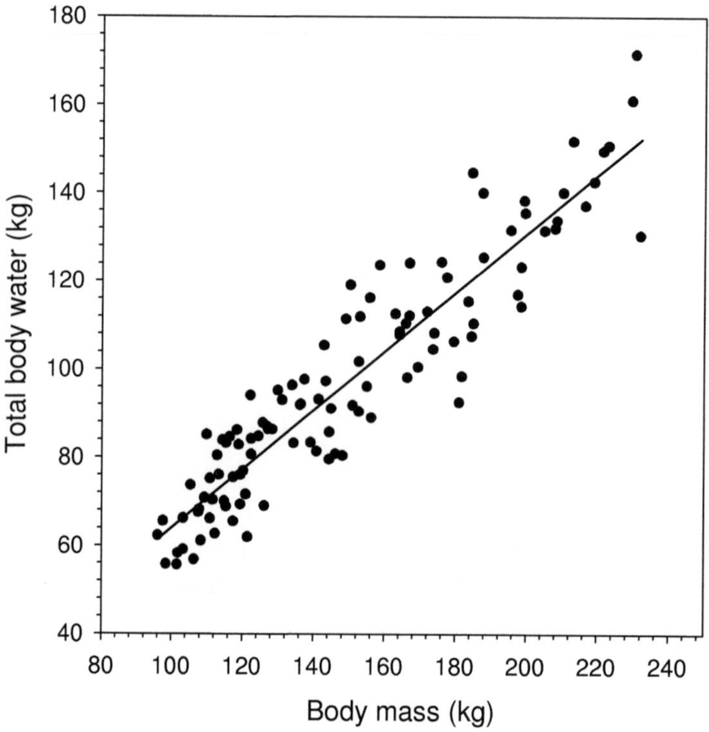

Fig 1. Relationship between body mass (kg) and total body water (TBW) in Shetland
pony mares (TBW, kg = -2.86 + 0.67·body mass, kg; R^2 = 0.86, P < 0.001,
n = 10).

The seasonal variation in mass-specific TWI followed a non-linear pattern with a
significant increase in summer and a decrease in autumn (TWI, mL·kg^{-1}·d^{-1} = 15.07 +
23.69·mo - 1.45·mo^2, R^2 = 0.64, P < 0.01; Figure 2). Highest and lowest mass-
specific TWI were recorded in September (135.1 ± 6.9 mL·kg^{-1}·d^{-1}) and January (50.5
± 19.9 mL·kg^{-1}·d^{-1}), respectively. While mass-specific TWI was highly correlated with
LA (r = 0.78, P < 0.01), heart rate (r = 0.74, P < 0.01) and body mass (r = 0.73, P <
0.01), the correlation with T_a was only moderate (r = 0.58, P < 0.05). However, mass-
specific TWI in January at T_a's < 0 °C (n = 7) deviated from the general linear trend
(Figure 3). After exclusion of these data, re-calculated correlations with mass-specific

TWI were considerably higher (LA: $r = 0.87$, $P < 0.001$; heart rate: $r = 0.88$, $P < 0.001$; T_a: $r = 0.86$, $P < 0.001$).

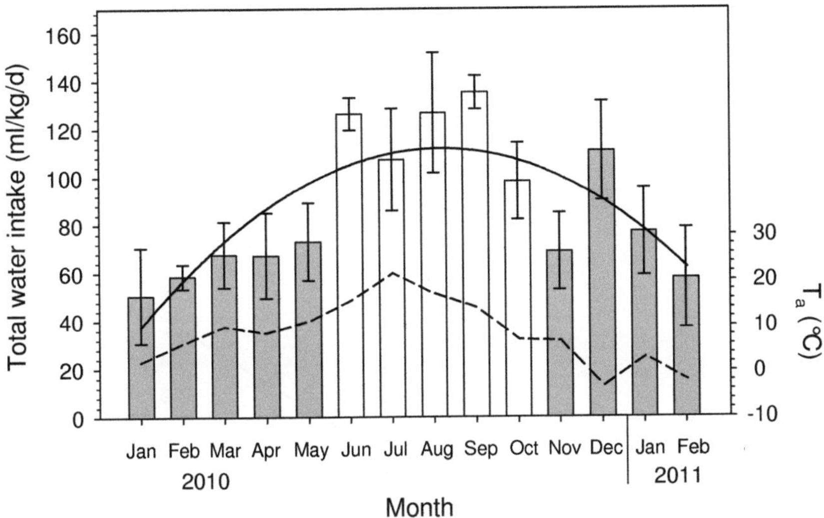

Fig 2. Seasonal changes in ambient temperature (T_a, dashed line) and total water intake (TWI, means ± SD) in Shetland pony mares (grey bars: open stable with paddock; white bars: pasture) estimated by the deuterium dilution technique. The solid line is described by the equation: TWI, $mL \cdot kg^{-1} \cdot d^{-1} = 15.07 + 23.69 \cdot mo - 1.45 \cdot mo^2$ ($R^2 = 0.64$, $P < 0.01$). Water was available *ad libitum* at all times.

The animal effect was highly significant, underlining large individual variation for mass-specific TWI ($P < 0.001$). The multiple regression calculated between mass-specific TWI, T_a and heart rate (TWI, $mL \cdot kg^{-1} \cdot d^{-1} = -13.38 + 1.77 \cdot$ heart rate, $beats \cdot min^{-1} + 2.11 \cdot T_a$, °C, $P < 0.001$) explained 84% of the variation in mass-specific TWI. The relationships between mass-specific TWI and heart rate (TWI, $mL \cdot kg^{-1} \cdot d^{-1} = -1.44 + 2.0 \cdot$ heart rate, $beats \cdot min^{-1}$, $R^2 = 0.40$, $P < 0.001$), and mass-specific TWI and T_a (TWI, $mL \cdot kg^{-1} \cdot d^{-1} = 66.34 + 2.69 \cdot T_a$, °C, $R^2 = 0.25$, $P < 0.001$) indicate increasing water intakes with an increase in both parameters. When mass-specific TWI values measured below 0°C were excluded, the coefficient of variation improved considerably for both relationships (TWI, mL·kg^{-1}·d^{-1} = 42.68 + 4.77·T$_a$, °C, R^2 = 0.54, P < 0.001; TWI, mL·kg^{-1}·d^{-1} = 21.8 + 2.35·heart rate, beats·min^{-1}, R^2 = 0.54, P < 0.001; Figure 3 and 4).

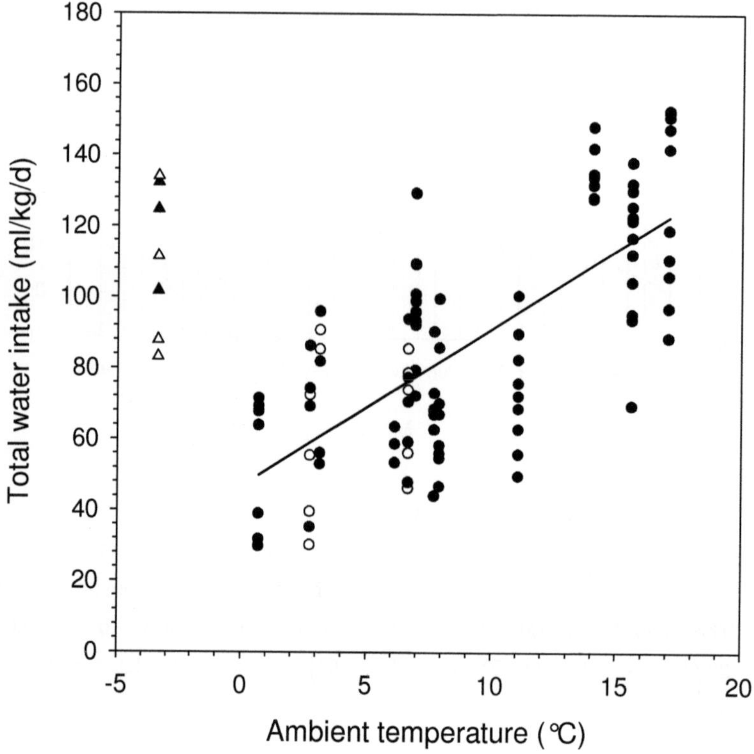

Fig 3. Relationship between total water intake (TWI) and ambient temperature (T$_a$) in restrictively fed (open symbols) and control fed (closed symbols) Shetland pony mares (TWI, mL·kg^{-1}·d^{-1} = 46.8 + 4.45·T$_a$, °C; R^2 = 0.52, P < 0.001, n = 10). Values measured at T$_a$ < 0°C (triangles) were excluded from the regression analysis (see text for details).

Discussion

Our study is the first measuring TBW und TWI in an herbivore over an extended period of time (13 mo) under temperate seasonal weather conditions. Out results suggest that TWI is higher for ponies than previously expected and that mass-specific TWI is strongly affected by T_a, locomotion and heart rate.

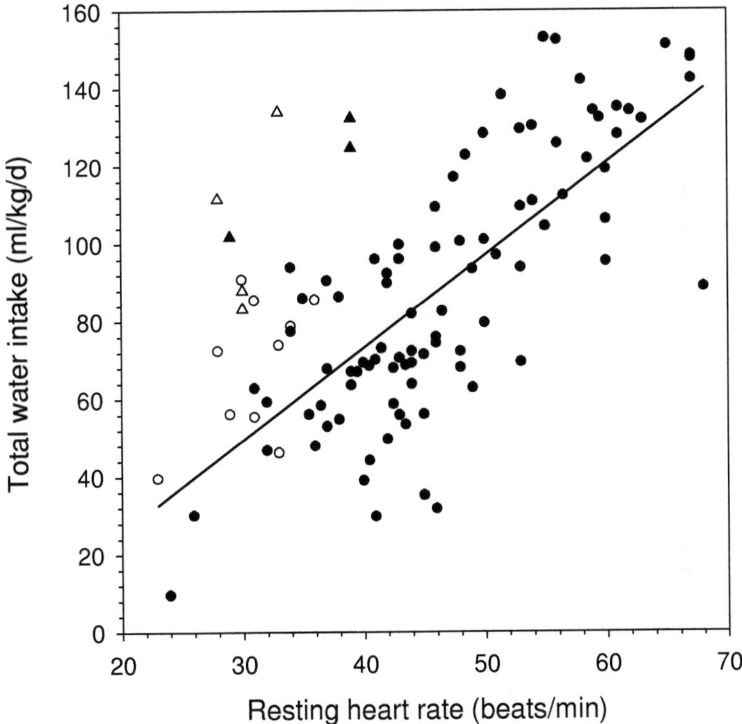

Fig 4. Relationship between total water intake (TWI) and resting heart rate (HR) in restrictively fed (open symbols) and control fed (closed symbols) Shetland pony mares (TWI, mL·kg^{-1}·d^{-1} = -21.81 + 2.35·resting heart rate, beats·min^{-1}; R^2 = 0.52, P < 0.001, n = 10). Triangles represent values measured at ambient temperatures < 0 °C (see text for details).

Total Body Water

In our study the TBW (kg) changed linearly with body mass. The resulting regression equation (TBW, kg = -2.86 + 0.67·body mass, kg) can be used to predict TBW contents for a large size range of ponies with high accuracy and provide e.g., information for medical treatments or prediction of body composition. Our prediction equation agrees with TBW (kg) contents for Namib horses, Quarter horses, Thoroughbreds, Percherons and ponies calculated by other authors (Julian et al., 1956; Deavers et al., 1973; Sneddon et al., 1991). Similar prediction equations have been derived, e.g. for cattle (Andrew et al., 1995) and llamas (Riek and Gerken, 2010). In contrast to TBW (kg), the TBW (%) remained fairly constant with changing body mass, indicating that the water fraction of the body was not influenced by the size of the animals. The measured TBW values ranging from 57.8 ± 1.8% to 71.2 ± 5.0% (mean 64.9 ± 4%) in our ponies were in the normal range reported for adult mammals (Hinchcliff et al., 1997; Toerien et al., 1999; Riek and Gerken, 2010, Al-Ramamneh et al., 2010, 2011). Our values are also in line with data on TBW in other equines ranging from 55 to 67% (Julian et al., 1956; Fielding et al., 2004; Radostits et al., 2005) and in horses and ponies (Deavers et al., 1973).

Total Water Intake

Our study is the first to report seasonal variations of TWI in horses on pasture based on the isotope dilution technique. While there are several studies measuring drinking water intake in horses (McDonnell and Kristula, 1996; Meyer et al. 1990; Crowell-Davis et al., 1985) none of them measured TWI, which includes not only drinking water but also preformed water in feed and metabolic water. Published experimental results indicate high variations in the intake of drinking water in horses. While the wild ancestor of the horse, the Przewalski horse, drinks on average about 10 to 60 mL·kg^{-1}·d^{-1} (Scheibe et al., 1998), domesticated horses are reported to consume 25 to 80 mL·kg^{-1}·d^{-1} (Meyer et al., 1990; Johnson, 1998). Our results on TWI based on the isotope dilution technique are with 50.5 to 135.1 mL·kg^{-1}·d^{-1} about 1.7 to 5.1 times higher than in the afore mentioned studies on drinking water. The difference between our TWI data and reported drinking water values for horses is thus approximately 40 mL·kg^{-1}·d^{-1}. Thus, a considerable amount of water ingested originates from other sources than drinking water. Assuming that the average water content of fresh grass is approximately 80%, the calculated difference between TWI and average drinking water intake in horses of 40 mL·kg^{-1}·d^{-1}, would amount to approximately 50 g grass intake·kg^{-1}·d^{-1} not considering water derived from metabolic water.

Our long term study revealed pronounced seasonal variations. In late summer our animals consumed considerably more water than in winter. Increased drinking water intake in summer was also found by McDonnell and Kristula (1996) in pony stallions (approx. 42 mL·kg^{-1}·d^{-1}). Increased T_a's in summer are expected to activate thermoregulative mechanisms for dissipation of metabolic heat to keep the body core temperature in a physiological range. As evaporation in association with sweat production are the main mechanisms of heat loss in horses under higher temperatures (Morgan, 1997; Hoffmann et al., 2009), large amounts of water may be lost with increasing T_a thus explaining the higher need for water intake. In winter we observed lower TWI in our ponies (January: 50.1 mL·kg^{-1}·d^{-1}) compared to summer confirming results from Kristula and McDonnell (1994). In a study on Welsh pony mares drinking water consumption approached 0 with T_a's below 10 °C (Crowell-Davis et al., 1985). Contrarily, our animals showed increased TWI with T_a's below 0 °C, irrespective of the feeding regime. Our study is the first reporting this kind of phenomena in a herbivore. A possible explanation might be that throughout the time T_a's were below 0 °C (December 2010) a closed snow cover of 1 to 23 cm was present and animals might have consumed snow while foraging. In addition, high dry matter contents in winter diets based on hay and straw lead to a reduced digestibility and an increased amount of feces. Thus, higher amounts of water are lost via feces and drinking water intake may have increased accordingly (Meyer, 1992). Another reason might be that the thick snow cover constrained foraging and led to higher drinking water intake as a compensating behavior. Alternatively, the amount of metabolic water generated may have increased in the context of chemical thermogenesis (Sileshi et al., 2002) to cover energetic demands under very cold temperatures.

A large amount of the observed seasonal variation in TWI (84%) was explained by changes in heart rate and T_a. After removal of the mass-specific TWI values measured at a T_a below 0 °C, the relationships between TWI and T_a as well as between TWI and heart rate were highly significant, explaining more than 54% of the variation in TWI. In this context, it is of interest to note, that only T_a and heart rate had significant effects on TWI, whereas the influence of LA was minor. Indeed, activity elevates the heart rate and simultaneously the thermoregulation needed to eliminate the heat produced by a higher muscle tonus (Kamen, 2001). However, the observed increase in TWI with increasing heart rate can not be attributed to higher LA of the animals, because we measured heart rate at rest. The simultaneous increase in heart rate may be attributed to an increase in metabolic activity due to thermoregulative processes, but also to higher digestive activity due to high food intake in summer (Arnold et al., 2006). Under our extensive free ranging conditions, however, both increases in heart rate and T_a were intermingled.

Both feeding groups did not differ in their TWI. Contrarily, Gupta et al. (1999) observed reduced drinking water intake in mules and donkeys during starvation and Kronfeld (1993) reported the same effect for horses. Although heart rate and body mass decreased significantly in our restricted animals, there was no significant impact on TWI. The restrictively fed group would have been expected to have lower TWI as they consumed less food. However, the comparable TWI to the control group may have been caused by the development of an over-drinking (polydipsia) behavior that is motivated by hunger as seen e.g. in pigs (Yang et al., 1984). An alternative explanation could be an increase in the generation of metabolic water due to metabolic processes such as protein and fat degradation to cover energy demands (Young, 1976; Milbury, 1997).

Our study is the first to report long term changes in TBW and TWI of ponies in an extensive outdoor housing system and shows that under year round extensive housing conditions, seasonal variations in water flux closely follow environmental conditions. We conclude that the TBW (%) in our horses was independent of T_a, LA, body mass and feed restriction. The calculated prediction equation for TBW (kg) can serve as an orientation e.g., for administrating drug doses. The TWI increased with increasing T_a, LA, heart rate, and mass. The comparison between reported data on drinking water intake and our TWI measurements indicate that horses on pasture ingest considerable amounts of water from other sources than drinking water (up to 70%). Particular attention needs to be paid to the water intake at temperatures below 0 °C as the water demand may be higher than expected. Contrary to expectation, water supply in winter might be more critical than under summer conditions, when the water content of grass is high. Thus, the adequate water supply of horses under year round extensive housing conditions requires special attention to meet maintenance and welfare requirements.

References

Al-Ramamneh, D., A. Riek, and M. Gerken (2010). Deuterium oxide dilution accuracy predicts water intake in sheep and goats. *Animal* 4, 1606-1612.

Al-Ramamneh, D., M. Gerken, and A. Riek (2011). Effect of shearing on water turnover and thermobiological variables in German blackhead mutton sheep. *J. Anim. Sci.* 89, 4294-4304.

Andrews, F. M., J. A. Nadeau, L. Saabye, and A. M. Saxton (1997). Measurement of total body water content in horses, using deuterium oxide dilution. Am. J. Vet. Res. 58, 1060-1064.

Arnold, W., T. Ruf, and R. Kuntz (2006). Seasonal adjustment of energy budget in a large wild mammal, the Przewalski horse (*Equus ferus przewalskii*) II. Energy expenditure. *J. Exp. Biol.* 209, 4566-4573.

Bakker, H. R., and A. R. Main (1980). Condition, body-composition and total-body water estimation in the quokka, *Setonix-Brachyurus* (Macropodidae). *Aust. J. Zool.* 28, 395-406.

Carroll, C. L., and P. J. Huntington (1988). Body condition scoring and weight estimation of horses. *Equine Vet. J.* 20, 41-45.

Crowell-Davis, S. L., K. A. Houpt, and J. Carnevale (1985). Feeding and drinking behavior of mares and foals with free access to pasture and water. *J. Anim. Sci.* 60, 883-889.

Deavers, S., Rosborou.Jp, H. E. Garner, R. A. Huggins, and J. F. Amend (1973). Blood volumes and total-body water in domestic pony. *J. Appl. Physiol.* 34, 341-343.

Fielding, C. L., K. G. Magdesian, D. A. Elliott, L. D. Cowgill, and G. P. Carlson (2004). Use of multifrequency bioelectrical impedance analysis for estimation of total body water and extracellular and intracellular fluid volumes in horses. *Am. J. Vet. Res.* 65, 320-326.

Forro, M., S. Cieslar, G. L. Ecker, A. Walzak, J. Hahn, and M. I. Lindinger (2000). Total body water and ECFV measured using bioelectrical impedance analysis and indicator dilution in horses. *J. Appl. Physiol.* 89, 663-671.

Gehre, M., H. Geilmann, J. Richter, R. A. Werner, and W. A. Brand (2004). Continuous flow $^2H/^1H$ and $^{18}O/^{16}O$ analysis of water samples with dual inlet precision. *Rapid Commun. Mass Spec.* 18, 2650-2660.

GfE - Gesellschaft für Ernährungsphysiologie (1994). Empfehlung zur Energie- und Nährstoffversorgung der Pferde. DLG-Verlag, Frankfurt a. M., Germany.

Gupta, A. K., Y. P. Mamta, and M. P. Yadav (1999). Effect of feed deprivation on biochemical indices in equids. *J. Equine Sci.* 10, 33-38.

Hinchcliff, K. W., G. A. Reinhart, J. R. Burr, C. J. Schreier, and R. A. Swenson (1997). Metabolizable energy intake and sustained energy expenditure of Alaskan sled dogs during heavy exertion in the cold. *Am. J. Vet. Res.* 58, 1457-1462.

Hoffmann, G., S. Rose-Meierhofer, and B. Niemann (2009). Water demand of horses and water consumption in horse husbandry. *Tierärztl. Umschau* 64, 438-442.

Hollemann, D. F., R. G. White, and J. R. Luick (1982). Application of the isotopic water method for measuring total body water, body composition and body water turnover. Pages 9-32 in Use of tritiated water in studies of production and adaptation in ruminants. Vienna, International Academic Energy Agency.

Johnson, P. J. 1998. Physiology of body fluids in the horse. *Vet. Clin. N. Am. Equine* 14, 1-22.

Julian, L. M., J. H. Lawrence, N. I. Berlin, and G. M. Hyde (1956). Blood volume, body water and body fat of the horse. *J. Appl. Physiol.* 8, 651-653.

Kamen, G. 2001. Foundation of Exercise Science. Lippincott Williams & Wilkins, Ambler, USA.

Kristula, M. A., and S. M. McDonnell (1994). Drinking-water temperature affects consumption of water during cold weather in ponies. *Appl. Anim. Behav. Sci.* 41, 155-160.

Kronfeld, D. S. (1993). Starvation and malnutrition of horses - recognition and treatment. *J. Equine Vet. Sci.* 13, 298-304.

Lifson, N., G. B. Gordon, and R. McClintock (1955). Measurement of total carbon dioxide production by means of D_2O. *J. Appl. Physiol.* 7, 704-710.

McDonnell, S. M., and M. A. Kristula (1996). No effect of drinking water temperature (ambient vs. chilled) on consumption of water during hot summer weather in ponies. *Appl. Anim. Behav. Sci.* 49, 159-163.

Meyer, H., Y. Gomda, H. P. Noriega, M. Heilemann, and A. Hipp-Quarton (1990). Investigations on the postprandial water metabolism in resting and exercised ponies. *Adv. Anim. Physiol. Anim. Nutr.* 21, 35-51.

Milbury, P.E. (1997). CEAS generation of large multi-parameter metabolic databases for determining categorical process involvement of biological molecules. Pages 127–144 in Coulometric electrode array detectors for HPLC. I. N. Acworth, M. Naoi, H. Parvez, and S. Parvez, eds. VSP International Science, Utrecht, The Netherlands.

Morgan, K. (1997). Dissipation of heat from standing horses exposed to ambient temperatures between -3 °C and 37 °C. *J. Therm. Biol.* 22, 177-186.

Oftedal, O. T., H. F. Hintz, and H. F. Schryver (1983). Lactation in the horse - milk-composition and intake by foals. *J. Nut.* 113, 2096-2106.

Radostits, O. M., C. C. Gay, D. C. Blood, and K. W. Hinchcliff (2005). Veterinary medicine - A textbook of the diseases of cattle, sheep, pigs and horses. 9th ed. W.B. Sauders, Philadelphia, USA.

Riek, A., and M. Gerken (2010). Estimating total body water content in suckling and lactating llamas *(Lama glama)* by isotope dilution. *Trop. Anim. Health Pro.* 42, 1189-1193.

Robb, J., J. T. Reid, M. S. S. Rhee, H. F. Schryver, H. F. Hintz, R. B. Harper, and J. E. Lowe (1972). Chemical composition and energy value of body fatty-acid composition of adipose-tissue, and liver and kidney size in horse. *Anim. Prod.* 14, 25-34.

SAS (2008). User's guide Release 9.02. SAS Inst., Cary, NC, USA.

Scheibe, K. M., K. Eichhorn, B. Kalz, W. J. Streich, and A. Scheibe (1998). Water consumption and watering behavior of Przewalski horses (*Equus ferus przewalskii*) in a semireserve. *Zoo Biol.* 17, 181-192.

Schoeller, D. A., E. Ravussin, Y. Schutz, K. J. Acheson, P. Baertschi, and E. Jequir (1986). Energy expenditure by doubly labeled water validation in humans and proposed calculation. *Am. J. Physiol.* 250, 823-830.

Sileshi, Z., A. Tegegne Tekle, and G. Tsadik (2002). Water resources for livestock in Ethiopia: Implications for research and development. Pages 66-79 in International Workshop Proceedings: MoWR/EARO/IWMI/ILRI. ILRI, Addis Ababa, Ethiopia

Sneddon, J. C., J. G. Vanderwalt, and G. Mitchell (1991). Water homeostasis in desert-dwelling horses. *J. Appl. Physiol.* 71, 112-117.

Toerien, C. A., T. Sahlu, and W. W. Wong (1999). Energy expenditure of Angora bucks in peak breeding season estimated with the doubly-labeled water technique. *J. Anim. Sci.* 77, 3096-3105.

Yang, T. S., M. A. Price, and F. X. Aherne (1984). The effect of level of feeding on water turnover in growing pigs. *Appl. Anim. Behav. Sci.* 12, 103-109.

Young, R.A. 1976. Fat, energy and mammalian survival. *Am. Zool.* 16, 699-710.

CHAPTER 6

GENERAL DISCUSSION

GENERAL DISCUSSION

Interspecies comparison

The present study was conducted to contribute to the basic knowledge on the adaptation capacity of domestic horses when kept under close to natural conditions. Accordingly, the experimental animals were exposed to high seasonal climatic fluctuations and in addition likewise to variations in food quality and quantity. The feed restriction of five ponies in the second winter was undertaken to simulate the natural feed scarcity that can be observed under very extensive housing conditions like e.g., in landscape management projects. However, feed shortage can also appear in semi-extensive housing by an inadequate feeding regime. The ponies adapted to the fluctuations in feed availability and environmental conditions by changes in physiological and behavioural reactions. The interspecies comparison allows pointing out differences and similarities in the reaction of herbivores to extensive conditions as in our experiments. In the following the focus is on a comparison of the adaptation strategies in different domestic herbivores.

Adaptation strategies of different herbivore species to feed shortage

Starvation or extended feed restriction does not only result in hunger and the increased motivation of feed intake for the animal but simultaneously affects behavioural, physiological and hormonal parameters to counteract these challenges (Broom and Fraser, 2007). In this section the influence of feed shortage and starvation on livestock species is analysed and compared to the present findings in horses. For a suitable comparison the focus was on the ruminating domestic herbivores cattle, sheep and goat.

Hunger and feed intake is subjected to a cross-linked system of neural, hormonal and metabolic processes (Porzig and Sambraus, 1991; Forbes, 2003). The regulation of feed intake can be subdivided in short-term and long-term regulation mechanisms. Short-term mechanisms regulate the alteration of the beginning and the end of feed intake, whereas the necessity for feed intake is kept up continuously but is regularly inhibited by different factors. Long-term mechanisms are related to the preservation of the energy balance and the energy reserves over a longer period (Porzig and Sambraus, 1991; Meyer, 1992).

The feeling of hunger is usually accompanied by contractions of the stomach. In herbivores this reaction already occurs before the stomach is completely empty. Horses show these contractions 5 hours following the last feed intake (Kolb, 1967) while in ruminants the reaction may appear later due to the large rumen volume.

In horses a continuous intake of small portions of feed is typical (Zeitler-Feicht, 2001) as the stomach is small and releases mash continuously to the gut. Feed is consumed until the hunger vanishes (Porzig and Sambraus, 1991) but it should be noted that horses possess no stretch receptors in the stomach (Meyer, 1992). Generally, horses need to consume huge amounts of crude fibre to reach satiation (Porzig and Sambraus, 1991). Ruminants, with their forstomach, can consume large amounts of feed in a short time (Porzig and Sambraus, 1991), though they possess stretch receptors in the forstomach (Scheunert and Trautmann, 1987). In ruminants the effect of fasting may be delayed due to the rumen volume that stores large amounts of feed (Heitmann et al., 1986).

The normal reaction of the animals' body to feed restriction, especially during seasons of high energy requirement, is the decrease in insulin secretion by decreasing glucose concentrations. This reaction was found in horses (Glade and Reimers, 1985), cattle (Ward et al., 1992), goats (Matsuyama et al., 2004) as well as in sheep (Heitmann et al., 1986) and results in an increased ratio of growth hormones to insulin and therefore in a decreased antilipolytic effect of insulin (Yang and Baldwin, 1973). Thus, the lipolytic hormones (epinephrine, norehinephrine, growth hormone) mobilise triglycerides form the horses' body fat. This leads to an increase in non-esterified fatty acids (NEFA) concentration in the plasma of the experimental ponies as seen in chapter 3. This reaction is also found in sheep, goats, larger horses and cattle (Rule et al., 1985; Heitmann et al., 1986; Sticker et al., 1995; Matsuyama et al., 2004).

Lactating cows react more sensible to starvation than other ruminants as shown by the higher increase in NEFA and faster decrease in glucose concentrations compared to sheep (Baetz, 1976). Obese horses are predisposed to insulin resistance resulting in high fat mobilisation rate during fasting, though ponies are more insulin insensitive than larger horses so that body fat is mobilised even faster during starvation (Dahme and Weiss, 2007).

NEFA are transported to the liver where they normally are introduced in the citric acid cycle and catabolised to energy. If the amount of NEFA mobilised exceeds the capacity of the citric acid cycle, the NEFA are converted to ketone bodies or re-esterified to triglycerides (von Engelhardt and Breves, 2000). In monogastric animals ketone bodies are synthesized exclusively in the liver, however, in horses only small amounts of ketone bodies are formed (Sjaastad et al., 2003). Instead, more NEFA is re-esterified to triglycerides leading to increased concentrations in plasma lipids and to fat accumulation in the tissues (von Engelhardt and Breves, 2000). Especially in ponies, fasting can increase the plasma triglyceride concentrations (Morris et al., 1972). Ruminants tend towards ketosis as NEFA is increasingly transformed to the ketone bodies β-hydroxybutyrate (BHB), acetone and acetoacetat in the liver (Katz and Bergman, 1969), thus increasing the ketone body concentration in the plasma of goats (Matsuyama et al., 2004) cattle (Nielsen et al., 2003) and sheep (Heitmann et al., 1986). However, cattle exhibited a distinct higher increase in ketone body concentrations during fasting than sheep (Baetz, 1976) and cows diseased with ketosis usually developed further illness (e.g., metritis, retained placenta) (Staufenbiel, 2004).

In ruminants BHB are likewise synthesized from butyrate in the epithelium of the forstomach. This transformation is important as butyrate inhibits the gluconeogenesis from propionate and therefore leads to a further increase in NEFA mobilisation and ketone body formation (Sjaastad et al., 2003) in restricted fed ruminants. Nevertheless, in unfed ruminants the liver becomes the sole organ of ketogenesis (Katz and Bergman, 1969).

The experimental ponies decreased BHB concentration during feed restriction, due to the decreased ketone body production in the liver, while in ruminants the BHB and acetone concentrations increased with fasting (Rule et al., 1985; Heitmann et al., 1986; Frohli and Blum, 1988). During starvation most tissues switch to lipids as their main energy source. This saves glucose for the brain that is reliant on the supply of the latter (von Engelhardt and Breves, 2000). In most species, like the horse, the brain can utilize glucose as well as ketone bodies as energy source when the glucose concentration falls (Sjaastad et al., 2003). In fasted sheep the use of ketone bodies by the brain is minimal compared with that of glucose (Pell and Bergman,

1983). Similar is known for other ruminants where ketone bodies do not represent an alternative source of energy as the brain is unable to limit the utilization of glucose (Sjaastad et al., 2003). This indicates the importance of normoglycemia in ruminants (Rule et al., 1985).

The high concentrations of NEFA, that need to be transformed in the liver, impair the esterification of the toxic unconjugated bilirubin to the non-toxic conjugated bilirubin in the liver, thus increasing the bilirubin concentration in the blood (Kraft and Dürr, 2005). Bilirubin concentrations in horses are usually higher than in other species (von Engelhardt and Breves, 2000). In the restricted ponies of the present experiment the bilirubin concentrations exceeded the normal range getting close to the concentration known to cause icterus. In cattle, goats and sheep a hepatic insufficiency due to fasting usually results in an elevated bilirubin concentration (Galyean et al., 1981; Ali et al., 1984), however, the increase may not be sufficient to result in icterus (Finn and Tennant, 1974; McSherry et al., 1984).

Fasting bulls can be characterized as metabolically hypoglycaemic (Trenkle, 1976; Sjaastad et al., 2003), hyperketonemic (Emmanuel and Kennelly, 1984; Sjaastad et al., 2003) and hyperlipidemic (Pothoven and Beitz, 1975). Similarly, fasting horses can develop hyperlipidemia (Morris et al., 1972; Seifi et al., 2002) and hyperbilirubinemia (Gronwall and Mia, 1972; Engelking, 1993), while sheep can become hyperbilirubinemic (Furll and Schafer, 1993) and hyperketonemic (Brockman, 1979).

Thyroxin (T_4) and triiodthyronin (T_3), hormones mainly responsible for the level of the basal metabolic rate, usually descend with decreased feed supply resulting in a lower metabolic rate and reduced energy requirements for maintenance (Ekpe and Christopherson, 2000). We could not found a significant difference in T_4 concentrations between our two feeding groups even if there was a significant decline in the 120 days lasting feeding trial in the restricted animals. In contrast, Blum and Kunz (1981) found that in cattle a five day starvation decreased the T_3 and T_4 concentrations. Fröhli and Blum (1988) only reported of decreasing T_3 concentrations in bulls during three days of fasting. In sheep decreasing T_3 levels were found after five weeks of feed restriction and a decrease in T_3 and T_4 after 18 weeks of caloric restriction (Blum et al., 1980; Ekpe and Christopherson, 2000). Abdullah and Falconer (1977) detected decreasing thyroid activity in goats fed restrictedly for five days.

The lower thyroid hormone concentrations are due to a lowered thyroid activity (Blincoe and Brody, 1955). Frequently observed constant T_4 values can be caused by a lower T_4 disappearance rate from the plasma as shown in cattle by Lundgren and Johnson (1964). Nevertheless, Iqbal (1990) found increased concentrations of T_3 and T_4 in seven days fasted goats. Indeed this would result in a fatal higher energy demand. According to Iqbal (1990) short term starvation decreases but long term starvation increases the T_3 and T_4 concentrations.

Beside changes in blood parameters the ponies showed changes in physiological and behavioural parameters (Chapter 2). Among these the heart rate was used to estimate the energy expenditure. The heart rate of the ponies decreased continuously during the four months of feed restriction from 35 to 26 beats/min. Reduced heart rate, usually being an indicator for a reduced metabolic rate, was also observed in cattle and goats subjected to feed restriction and starvation (Drivers and Peek, 2008; Puchala et al., 2009). Cattle having a normal resting heart rate of 48 – 80 beats/min decreased their heart rate below 40 beats/min during a two day fast (Clabough and Swanson, 1989). Feed restriction is, beside heart rate reduction, closely associated with bradycardia in cattle appearing after 2 days of fasting (Bednarski and McGuirk, 1986). Bradycardia can be attributed to the reduced ruminoreticular fill that results in slowing of heart rate by an increase in parasympathetic tone (Clabough and Swanson, 1989). Goats reduced their heart rate from a mean of 100 beats/min during *ad libitum* feeding to 52 beats/min during a four day fast (Puchala et al., 2009). A 12 day starvation of sheep also reduced the heart rate (Naqvi and Rai, 1991). It can therefore be suggested that all animals reduced their metabolic rate during feed shortage or starvation.

Another indicator for reduced metabolic rate is the decline in rectal temperature. Epke and Christopherson (2000) showed that sheep reduced their rectal temperature during long term feed restriction under cold conditions. This may indicate a lower metabolic response resulting in a lower temperature gradient between animal and environment, thus reducing the heat loss while the reduced body core temperature limits the tissue metabolism. The rectal temperature in sheep was also positively correlated with the metabolic rate (Ekpe and Christopherson, 2000). Naqvi and Rai (1991) showed reduced rectal temperatures during a 12 day fast in sheep. Cattle and goats exhibited lower rectal temperatures after 4 and 12 days of feed deprivation, respectively (Rumsey and Bond, 1976; Naqvi and Rai, 1991).

A reduced feed intake usually leads to a lower need of drinking water intake due to a lower quantity of ingested dry matter. Furthermore, a reduced metabolic rate, indicated by the reduced heart rates and rectal temperatures, and lower evaporative water losses in winter decrease the water demand. However, as shown in chapter 5 the restricted ponies in the study had no significantly lower total water intakes than the non-restricted animals in the winter month (10.9 vs. 12.0 litre, respectively, p>0.05). In dairy cattle Dahlborn et al. (1995) found a reduction of 70% in drinking water intake when the animals were feed deprived for 58 hours. Similarly, Bond (1976) and Clabough and Swanson (1989) found a remarkable reduction in water intake in beef cattle and dairy cows that were feed deprived for up to two days. Sheep fasted for 36 hours had a lower desire for water (NRC, 1964).

Domestic ruminants (sheep, goat, cattle) and horses show similar adaptive capacities by a reduction in metabolic rate, energy conservation by reduced activity and alternative provision of energy by mobilization of body fat reserves. However, there are remarkable species differences with regard to their physiological pathways. In particular, the horse appears to differ from the mentioned ruminants in insulin sensitivity, independence on glucose provision and the conversion of free fatty acids.

Water intake of domesticated herbivores

Water is the largest body constituent in animals and the most important nutrient. The regulations of body temperature, digestion, metabolism, excretion, growth and reproduction as well as the regulation of the mineral homeostasis and hydrolysis of proteins, fat, and carbohydrates are important processes which are dependant on the water supply (NRC, 2000). In environments with a limited water supply, quantitative information on the water intake of the farm animals is important (Winchester and Morris, 1956) to ensure sufficient watering and therewith health, performance and welfare for the animals (Bond et al., 1976). Surprisingly, little is known about the actual water requirements of livestock, presumably due to assumed abundance of water supply (Beede and Collier, 1986). Water requirement is influenced by several factors, including breed, age, physiological stage, body weight, environmental temperature and humidity, activity, lactation, rate and composition of gain, dry matter of feed, feed intake and water temperature (Murphy, 1992; Kristula and McDonnell,

1994; NRC, 2000; Coenen, 2005; Hoffmann et al., 2009). Differences in water requirement and tolerance of water shortage are to be expected between monogastric animals and ruminants as the latter possess the rumen as a large fluid reservoir which can store high amounts of water that can be used when water is a restricted resource (Silanikove and Tadmor, 1989).

Water is lost from the body by sweat, urine, faeces, milk and respiration (Looper and Waldner, 2007). All water lost needs to be compensated by water intake as the total body water of the animals must remain relatively constant (Macfarlane and Howard, 1972). Water requirements of livestock are satisfied by three major sources: water contained in feed, drinking water and metabolic water produced by the oxidation of organic nutrients. The catabolism of 1 kg fat, carbohydrate and protein offers 1190 g, 560 g and 450 g of water, respectively. This metabolic water can be important in periods of negative energy balance and in arid areas (NRC, 1981).

In chapter 5 of the present study we measured the total water intake of ponies which comprises all three water sources mentioned above. Since there are hardly any data available on total water intake in cattle, sheep and goats, data dealing with drinking water intake will also be used in the following section. However, a direct comparison is not possible as total water intake values are always higher than drinking water intake values.

We found highly varying total water intake rates in our ponies over the course of the year whith much higher total water intake in summer than in winter. Highest total water intake was observed in September with 135 ml/kg body weight (BW) (22.3 l/day) at a mean ambient temperature of 14 °C and lowest in January with 51 ml/kg BW (7.3 l/day) at mean ambient temperatures of 1 °C.

Similarly, the daily total water intake in sheep and goats was lower in winter (10-22 °C; 184 ml/kg BW and 131 ml/kg BW, respectively) in comparison with summer (29-44 °C; 300 ml/kg BW and 331 ml/kg BW, respectively) (Alamer, 2011). However, this total water intake measured in summer and winter for goats and sheep by Alamer (2011) was much higher than the total water intake of our ponies in these seasons. Sheep and goats, being a species adapted to arid environments, would have been expected to consume less water than species mainly adapted to temperate regions (horses, cattle). The total water intake rates found by Al Ramamneh et al. (2010) for goats (46.9 ml/kg BW) and sheep (79.7 ml/kg BW) at 14 °C and for sheep (50 ml/kg BW) at 20 °C seem to be more appropriate for animals adapted to arid conditions.

Aganga (1989) also found lower total water intake rates of 87 ml/kg BW in sheep and of 68 ml/kg BW in goats during temperatures ranging from 21 to 39 °C. Total water intake of sheep under hot summer conditions found by MacFarlane (1966) was in the same range (81.5 ml/kg BW). Non lactating cows (Holstein) showed a total water intake of 73 ml/kg BW at an ambient temperature of 18 °C and beef cattle (Simford) of 157 ml/kg BW at temperatures of 38 °C.

A reduced intake of drinking water was also registered for feedlot cattle in winter with 19.0 l/day compared to summer water intake of 31.2 l/day (Hoffman and Self, 1972). Dairy cows consumed 38.6 l drinking water/day in winter and 61.9 l drinking water/day in summer (Lainez and Hsia, 2004). Kamal (1959) measured a significantly higher amount of water drunk by cattle exposed to 35 °C compared to 2 °C (67 and 23 litres, respectively).

Based on the lower water content in faeces, sheep and goat (40-65% water content of faeces) generally need to consume less water than cattle (75-85% water content of faeces) and horses (about 75% water content of faeces) (NRC, 1981; Aganga et al., 1989; Meyer, 1992). Furthermore, the ability to concentrate the urine is higher in sheep and goats (Maloiy, 1973). Especially goats are acclimatised to arid regions and dehydration because of their small body size, low metabolic requirements and the ability to minimize water losses via urine and faeces (Silanikove, 2000).

Higher water intake in summer is mainly due to higher ambient temperatures. In the present study the ambient temperature had the highest effect on the total water intake. According to McDowell (1972) increased water consumption is a major response to thermal stress. Consumed water increases the animals comfort by a direct cooling of the reticulorumen (Bianca, 1963) and the physical properties of water are important for the transfer of heat from the body to the environment (Beede and Collier, 1986; Looper and Waldner, 2007). For horses as well as sheep, goats and cattle, the evaporative heat loss is the most important heat dissipation mechanism during high temperatures (Scheunert and Trautmann, 1987). Nevertheless, Mc Dowell (1972) reported high differences in the importance of sweating with domestic livestock ranked in a descending order of horses, donkeys, cattle, goats and sheep. The associated water loss may therefore be higher in equids.

During hot temperatures the increased evaporative cooling of the animals' body increases water losses (Kadzere et al., 2002). Non-lactating cattle increased their total water loss from the body by 58% at 30 °C compared with 20 °C. The main increase was due to a 176% increased sweat loss (McDowell and Weldy, 1967). Comparable results were noted for lactating cows with an increase in skin, surface and respiratory evaporation of 15, 59 and 50% (McDowell, 1972). Horses keep their evaporative heat loss constant below 20 °C, but a pronounced increase in evaporation was shown above 20 °C. The high evaporative heat loss is considered to be a short term effect (Morgan et al., 1997). Generally, the acclimatisation to heat, with decreased sweat flow and a decrease of salt content, takes four to six weeks (Guyton, 1991). However, the present ponies showed increased water intake over the entire summer period suggesting an inadequate adaptation of water use. The ambient temperatures increased continuously until summer, so that probability no complete adaptation was possible. In addition, the ponies increased their body condition score (BCS) over the summer and possibly these fat depots isolated the body (Guyton, 1991) thus hindering adequate thermoregulative mechanisms.

As shown in chapter 3 the ponies increased their T_4 concentrations in summer compared to winter. The associated higher metabolism in summer increases the drinking water demand and additionally raises the water produced by metabolic reactions what leads to higher total water intake rates. Similarly, higher metabolic rates in sheep increased their drinking water intake (NRC, 1985). In cows the metabolism was reduced during heat stress in order to reduce the metabolic heat needed to be lost from the body (Christopherson and Kennedy, 1983). Also Abdullah and Falconer (1977) found reduced thyroid activities in goats exposed to heat stress. In the present experiment the ambient temperatures may not have been high enough to cause a decreased metabolism in summer due to heat stress. Ott (2005) stated that increased body core temperatures, sweating and metabolic rates with higher temperatures only appear before the onset of acclimation and that these effects appear to be transitory.

In our study the locomotor activity was highly correlated with the water intake but had no significant effect upon it. Activity will increase the water intake in all grazing animals (Vallentine, 2001) as the water loss through sweating and evaporation increases (NRC, 2000). However, there were no data available for water intake during activity for cattle, sheep and goats.

To ensure the water supply of livestock animals under different feeding regimes and ambient temperatures, prediction equations were developed, including different prediction equations for the drinking water intake of lactating cows (Castle and Thomas, 1975; Beede, 1993; Dahlborn et al., 1998; Looper and Waldner, 2007). However, the comparison of lactating and non-lactating animals is not suitable, so that these prediction equations will be discarded for the further discussion. For the prediction of drinking water intake in cattle, two equations were published. Similar equations for sheep and goats are lacking. Looper and Waldner (2007) presented a prediction equation for non-lactating cows considering the concentration of dry matter in the diet, dry matter intake and amount of protein in the diet.

Drinking water intake, kg/day =   - 10.34
$$+ \ 0.23 \ x \ \text{Diet dry matter (\%)}$$
$$+ \ 0.45 \ x \ \text{Dry matter intake (kg/day)}$$
$$+ \ 0.04 \ x \ \text{Diet crude protein (\%)}^2$$

This equation does not include the ambient temperature, an important effect on water intake.

A prediction equation for the drinking water intake of growing bulls was developed by Meyer et al. (2006):

Drinking water intake, kg/day =   - 3.85
$$+ \ 0.51 \ x \ \text{average ambient temperature}$$
$$+ \ 1.49 \ x \ \text{dry matter intake (kg/day)}$$
$$- \ 0.14 \ x \ \text{roughage part of the diet (\%)}$$
$$+ \ 0.25 \ x \ \text{dry matter content of roughage (\%)}$$
$$+ \ 0.014 \ x \ \text{body weight (kg)}$$

The only prediction equation found for prediction of total water intake in non-lactating cows was developed by Holter and Urban (1992). However, the total water intake was difined as free water intake and feed moisture:

Total water intake (kg/day) =   35.19

+ 0.98 x free water intake

- 0.01 x BW

+ 1.08 x dry matter intake (kg/day)

+ 1.18 x dietary crude protein (% of dry matter)

- 0.04 x dietary crude protein (% of dry matter)^2

- 0.99 x dietary dry matter (%)

+ 0.01 x dietary dry matter (%)^2

Such predictions depending on the exact information of the feed composition are not applicable to horses kept on the pasture for the whole year as composition and water content of the feed changes with season and intensity of grazing (NRC, 1981). We therefore designed a prediction equation only depending on the ambient temperature that is easily to measure. Our prediction equation for the total water intake on pasture is:

Total water intake, ml/kg BW day^{-1}=

46.76 + 4.45 x mean daily ambient temperature (°C)

There is only little information on the amount of water intake during different ambient temperatures for horses, cattle, sheep and goats as most studies calculated the water intake with the help of prediction equations or did not measure the ambient temperature during their experiments. All published data on total water intake data of sheep, goats and beef cattle at different temperatures are depicted in Fig. 1. The total water intake data of our poniers seem to fit well into the published data.

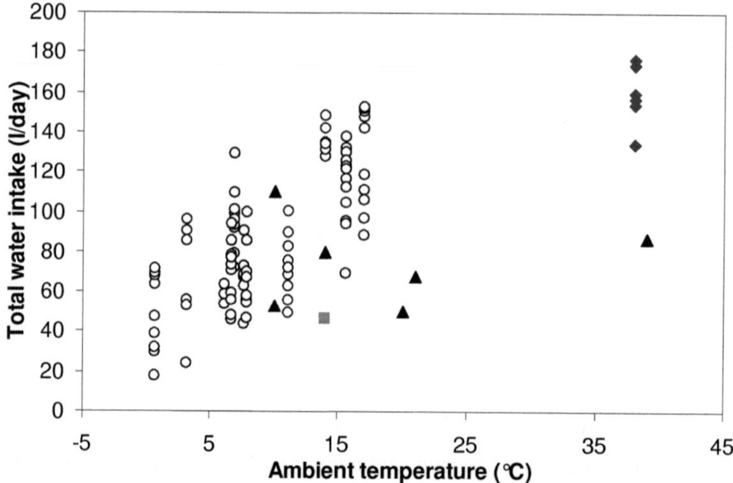

Fig. 1 Total water intake rates measured for sheep ▲, goats ■, beef cattle ♦ and the ponies of the present study ○ at different ambient temperatures (Al-Ramamneh et al., 2010; Aganga et al., 1989; Cassida et al., 1994; Silanikove and Tadmor, 1989).

In summary, the water intake of all described herbivores is influenced by similar parameters (ambient temperature, activity, feed quantity, feed composition, body fat). Cattle has the highest water demand per kg BW that can be mainly ascribed to the higher water content in the faeces. As expected, in lactating cows the higher demand is even more distinct due to the high water loss by the milk yield. Sheep and goats usually have a lower water demand especially at high ambient temperatures as they are able to reduce the volume and increase the concentration of their urine. Furthermore, they produce drier faeces than the other herbivores. With regard to efficient use of water, the compared species are tentatively ranked in the descending order of cattle, horses, sheep and goats.

Nocturnal hypometabolism

The results presented in chapter 2 suggest the existence of a hypometabolism in ponies. This is the first evidence of a hypometabolism in larger domesticated animals. Until now, similar mechanisms were only described in wild large Northern temperate mammals. Signer et al. (2011) detected a winter hypometabolism in Alpine ibex, Arnold et al. (2006) in Przewalski horses and Arnold et al. (2004) in red deer. In Tab. 1 the relevant results found in ponies are compared with those found in Przewalski horses, ibex and red deer.

In the studies of Arnold et al. (2004; 2006) and Signer et al. (2011) the animals were housed under extensive conditions while the ponies in the present study were kept under semi-extensive conditions. The Przewalski horses and the ibex were exclusively feeding on natural vegetation, the red deer and the ponies were fed supplementary in winter. However, five of the experimental ponies were fed restrictedly and five received feed *ad libitum.*

All studies measured the locomotor activity (LA), heart rate (HR) and body temperature of the animals over the entire year.

Similar seasonal activity pattern were observed in ibex, red deer, Przewalski horses and ponies with low activity in winter and increased activity in spring and summer. As shown in Tab. 1 the Przewalski horses present an activity peak in spring with a distinct reduction during summer while the other species were mostly active in summer. Parallel to the LA, the HR of the red deer, ibex, ponies and Przewalski horses increased rapidly in spring. In red deer and ibex the HR peaked in June and slowly declined thereafter to 40% of the yearly maximum in winter (Arnold et al., 2004; Signer et al., 2011). The ponies showed highest HR values in September followed by a rapid decline until winter by 50%. In the Przewalski horses the HR peaked earlier in May and remained on a medium level over summer and declined to a winter nadir of about 50% of the annual maximum (Arnold et al., 2006). Generally, the annual activity and heart rate rhythms of the described animals were similar. The temporal shift in seasonal patterns may be due to differences in the environmental conditions.

The observed annual and diurnal changes of HR in ibex, red deer, Przewalski horses and the experimental ponies of this study reflect changes in the metabolic rate as there is a close relation between HR and oxygen consumption (Renecker and Hudson, 1985; Bevan et al., 1995; Butler et al., 2004).

Tab. 1 Comparison of the relevant data found in the three studies on hypometabolism in large Northern ungulates

Species	Ibex (Capra ibex ibex)	Przewalski horses (Equus ferus przewalskii)	Red deer (Cervus elaphus)	Shetland ponies (Equus ferus caballus)	
Authors	Signer et al. (2010)	Arnold et al. (2006)	Arnold et al. (2004)	present study	present study
Housing conditions and measured parameters					
Animals (N)	10	7	9	5	5
Sex	5 males, 5 females	1 male, 6 females	4 males, 6 females	5 females	5 females
Housing	Extensive	Extensive	Extensive	Semi-extensive	Semi-extensive
Feeding	Natural forage	Natural forage	Natural forage, supplemented in winter	Natural forage in summer, ad libitum supplemented in winter	Natural forage in summer, restrively supplemented in winter
Measured parameters	Continuously: Locomotor activity (% of day), rumen temperature (°C), heart rate (beats/min), ambient temperature (°C), wind speed (m/sec), snow height (cm)	Continuously: Locomotor activity (% of day), subcutaneous temperature (°C), heart rate (beats/min), ambient temperature (°C), wind speed (m/sec)	Continuously: Locomotor activity (% of day), subcutaneous temperature (°C), heart rate (beats/min), ambient temperature (°C), wind speed (m/sec), humidity (%)	Both feeding groups Continuously: Locomotor activity (activity impulses), resting time (min/day), subcutaneous temperature (°C), ambient temperature (°C), precipitation (mm/cm) and humidity (%) Biweekly: resting heart rate (beats/min)	

Tab. 1 continued

Species	Ibex *(Capra ibex ibex)*	Przewalski horses *(Equus ferus przewalskii)*	Red deer *(Cervus elaphus)*	Shetland ponies *(Equus ferus caballus)*	
Daily rhythm					
Subcutaneous/rumen temperature	Nadir in rumen temperature around sunrise and peak in the afternoon	Lowest subcutaneous temperatures in early morning	Subcutaneous temperature decreased during night with nadir in the morning	Lowest subcutaneous temperatures around sunrise and peak in the afternoon	Lowest subcutaneous temperatures around sunrise and peak in the afternoon, higher fluctuations with reduced minimum values during feed restriction
Hearte rate	Lowest heart rates around sunrise and peak in the afternoon	-	-	-	-
Locomotor activity	Peak during day light hours, reduction during night	-	-	Nadir before sunrise, peak in early morning and around sunset	Nadir before sunrise, peak in early morning and around sunset

165

Tab. 1 continued

Species	Ibex (Capra ibex ibex)	Przewalski horses (Equus ferus przewalskii)	Red deer (Cervus elaphus)	Shetland ponies (Equus ferus caballus)
Annual rhytm				
Subcutaneous/rumen temperature	Highest values in late summer and nadir in spring	Low variations in daily mean and max Ts, considerable variations in daily min Ts with lowest values (24 °C) in April	Low variations in daily mean and max Ts, considerable variations in daily min Ts with lowest min Ts (16-17 °C) in February, highest (35 °C) in November	Low daily mean and max Ts variations, considerable variations in daily min Ts with lowest values in April (28 °C), highest (36 °C) in October	Low daily mean and max Ts variations, considerable variations in daily min Ts, lowest values in April (28 °C), highest (36 °C) in October, further reduction in mean and min Ts during feed restriction
Hearte rate	Rapid increase in spring, peak in June, reduction of 60% to winter nadir	Peak in May, medium level over summer, reduction in autumn to 50% of the annual maxium in winter	Peak in early June, reduction of 40% until nadir in late winter	Rise in spring and summer, peak in September, decline of 45% until November	Rise in spring and summer, peak in September, decline of 60% until February during feed restriction
Locomotor activity	Rise in spring, peak in summer, reduction in autumn until winter nadir	Increase in spring, peak in May, decline until winter nadir	Rise in spring, remains high during summer, reduction in autumn	Rise in spring, peak in summer and reduction in autumn, nadir in winter	Rise in spring, peak in summer and reduction in autumn, nadir in winter

In resting animals the relation between HR and metabolic rate is typically linear but curvilinear in exercising animals (Renecker and Hudson, 1985; Bevan et al., 1995; Brosh et al., 1998). The changes in heart rate may therefore not indicate changes in the metabolic rate but may rather be caused by changes in LA. In ibex, Przewalski horses and red deer the LA only partly explained HR changes. Furthermore, Signer et al. (2011) and Arnold et al. (2004; 2006) were able to measure the HR of ibex, Przewalski horses and red deer at rest, where the same changes accord as during activity. In ponies only the resting HR was measured, being marginally influenced by the total LA. Thus, the reduced metabolic rate found in ibex, red deer, Przewalski horses and ponies cannot be only attributed to a reduction in LA as indicated by the concomitant HR variations during rest.

Reduced metabolic rates in winter have also been reported for other wild Northern ungulates (Silver et al., 1969; Renecker and Hudson, 1985). Usually, contrary reactions are known to occur in metabolic rates during cold exposure with sufficient food supply (Sjaastad et al., 2003) in domestic species such as beef cattle, horses and sheep as shown by their increased thyroid hormone concentrations (Irvine, 1967; Westra and Christopherson, 1976; Christopherson et al., 1979). However, under these extensive winter conditions an increase of the metabolic rate would be fatal as the feed availability is low. Reduced metabolic rates in winter and the resumption of a high metabolic rate in spring as seen in the described studies are well known for hibernators and species undergoing daily torpor. The reductions in metabolic rate under these physiological states (torpor and hibernation) are much higher and mainly achieved by a decreased heat production and the tolerance of a decrease in body temperature (Geiser and Ruf, 1995).

In red deer, Przewalski horses and ponies the subcutaneous temperature was measured while in ibex the rumen temperature was determined. Mean and max daily subcutaneous temperatures showed little variations, while in contrast the daily minimum subcutaneous temperature varied considerably over the year in red deer, Przewalski horses and ponies with highest fluctuations in spring and lowest in early winter. In red deer the lowest daily minimum subcutaneous temperature was 16-17 °C, in Przewalski horses 24 °C and in ponies 28 °C.

Lower subcutaneous temperatures during low ambient temperatures may be attributed to physical thermoregulation. However, comparable but lower fluctuations were seen in daily mean rumen temperature of the ibex which cannot be caused by peripheral cooling. The rumen temperature stayed on a higher level, nevertheless, also showing lowest values in spring and highest in late summer. The reduced body temperatures in winter and spring may rather be due to a reduced temperature set-point and lower endogenous heat production by the down-regulated metabolism (Arnold et al., 2004; Signer et al., 2011). Similar decreases in the set-point can be noticed during the entrance into hibernation (Arnold et al., 2004).

In all studies the subcutaneous temperature was lowered during the night to a nadir in the early morning and subsequently increased rapidly. This pattern was paralleled by the ambient temperature so that it might be assumed that the skin temperature decreases due to peripheral cooling, however, the body temperature in our animals even decreased on days when the ambient temperature stayed relatively constant. In Przewalski horses, ponies and red deer the lowest daily subcutaneous temperatures appeared in spring some time after the lowest ambient temperatures. Even in summer when no physical thermoregulation to cold is necessary, variations in the subcutaneous temperature appeared in Przewalski horses, ponies and red deer. This phenomenon was attributed to a strategy to reduce the metabolic costs for thermoregulation in summer. The larger decrease in body temperature over night in summer increases the capacity to store heat during hot days (Arnold et al., 2006).

In our experiment restricted as well as *ad libitum* fed control animals showed reduced heart rates and T_4 concentrations in winter, whereas the reduction in the restricted animals was more intense. Furthermore, the subcutaneous temperature varied highly in late winter indicating a nocturnal hypometabolism. The restricted animals had significant lower subcutaneous temperatures on 52 of 120 day of feed restriction than the control animals and showed a further decrease in heart rate and T_4 concentrations indicating higher energy savings.

The described mechanisms in Alpine ibex, red deer and Przewalski horses are similar to that found in our ponies showing energy saving mechanisms in winter and summer. However, our observations indicate that the nocturnal hypometabolism in winter is also shown with adequate feed supply but the phenomenon is intensified by the feed scarcity in the restricted animals. Furthermore, there are similar mechanisms in domestic herbivores than in wild species when kept under extensive conditions at the Northern hemisphere.

The year round outdoor housing of horses under welfare aspects

In the following the reactions shown by the experimental ponies in chapter 2 and 3 will be used to determine if the extensive outdoor housing is an appropriate housing system for horses or if additional regulations need to be defined to prevent possible welfare impairment.

An increasing number of horses is kept in alternative housing systems including the year round outdoor housing (TVT, 2004). Until now there are no specific directives for the housing of horses in Germany apart from the Animal Welfare Act (BMELV, 2006) that only determines general requirements for animals. The German Ministry of Feed, Agriculture and Consumer Protection (BMELV) provides guidelines for the housing of horses, however, these guidelines do not specify the requirements for horses housed outdoor all year round (BMELV, 2009). The German veterinary office "Ennepe-Ruhr-Kreis" (Veterinäramt Ennepe-Ruhr-Kreis, 2011) developed recommendations on the basis of the Animal Welfare Act and the guidelines of the BMELV. According to these recommendations the year round outdoor housing is only an alternative for housing in stables if the following minimum requirements for outdoor group housing of horses are fulfilled: sufficient water and feed supply, possibility for the formation of winter coat, a weather protection with a dry and isolating ground (Ennepe-Ruhr-Kreis, 2011). These recommendations correspond with the guidelines in Switzerland (BVET, 2011) and Britain (DEFRA, 2009) and the statutory minimum standards in Austria (BMG, 2010). All these requirements were fulfilled in the housing of our control group. However, the restricted group received an insufficient amount of feed to simulate consequences of feed shortage in winter and to determine adaptation strategies to these conditions.

Low ambient temperatures increase the energy demand of animals (Autio, 2008) and with inadequate feed energy supply will result in weight loss of the animal due to the use of energy reserves to maintain the body temperature. The daily visual inspection of the horses is a prerequisite e.g. to ensure a sufficient feed supply (BMELV, 2009). However, weight loss up to 20% in the feed restricted ponies with winter coat was hardly recognizable by a visual inspection. Without touching the animal the body condition may therefore be misjudged and can result in further unintended insufficient feed supply, weight loss, reduced immunity and an impairment of health. This process takes place slowly and is therefore recognized rarely or too late by the owner

(Veterinäramt Ennepe-Ruhr-Kreis, 2011). Thus, visual inspection alone is not sufficient. An additional determination of the body weight or condition is necessary. Since weighing of the animals is usually impossible under extensive conditions an additional weekly body condition scoring is recommendable to detect changes in the animal condition at an early stage. However, in ponies the normal BCS may not be sufficient to detect changes in the body condition. A developed specific fat score system for the neck of ponies, the "cresty neck score", can provide additional information. In the Swiss guidelines for outdoor housing of horses the daily inspection is defined more precisely. The horse needs to be checked for injuries, parasites and excessive loss of weight (BVET, 2011). Such a specification should be included to the German guidelines.

Behaviour is one of the measures of welfare that is easily to observe. The comparison of free living and captive animals can provide information which behaviour is not shown in captivity but might be of importance for the animal to perform (Mench and Manson, 1997). The needs of our domesticated horse correspond largely with those of wild horses (Ministry of Food, 2009) as none of the wild horse behavioural patterns is lost in domesticated horses and only modifications of the behavioural patterns occurred during domestication (Zeitler-Feicht, 2001). According to Keeling and Jensen (2002) the normal behaviour is the behaviour that was developed during the evolutionary adaptation in the natural habitat.

The ponies in the present study, housed under semi-extensive conditions, showed a circadian and a diurnal rhythm in locomotor activity and resting time. The circadian rhythm wassimilar to that observed in Przewalski horses in a close-to-natural housing under comparable climatic conditions (Arnold et al., 2006; Berger et al., 2006) and in Camargue horses under extensive housing (Duncan, 1980). Likewise, the ponies' diurnal rhythm of activity and rest coincided with those found in Przewalski horses and free ranging Haflingers (Boyd, 1988; Berger et al., 1999; Lamoot and Hoffmann, 2004). The domestication did not change the natural time budget in robust horses and the year round outdoor housing of the present study enabled the performance of the natural activity and resting behaviour being important for the animals' welfare. However, the activity and resting time did not differ between the feeding groups even though it can be assumed that the welfare of the feed restricted animals was impaired, e.g. by the feeling of hunger (TVT, 2006). Thus, these results put the suitability of behaviour as an indicator for long term stress and welfare into question and also show that the visual control of the animals is insufficient. Accordingly, a combination of different indicators is necessary to adequately assess the animal welfare.

The feed restricted group had only access to feed during short periods of the day. This may result in frustration as the animal knows how to control the interaction with the environment but is prevented from carrying out the action (e.g., by feeding). Frustration results in abnormalities of physiology and behaviour indicating poor welfare (Broom, 1991).

The effective reaction to environmental challenges is essential for survival. When challenged, animals react by behavioural responses, but simultaneously physiological changes may occur (Moberg and Mench, 2000). Hormone concentrations in the plasma and cellular metabolic processes as well as the cardiac output can change (Willen, 2004). By the analyses of physiological indicators reactions of the animals can be detected that would be missed out by only ethological observation of the animal (Fell and Shutt, 1989). Long term stress, however, is difficult to detect as the animal possesses efficient feedback mechanisms that normalize the physiological parameters during prolonged stress (Terlouw et al., 1997; Keeling and Jensen, 2002).

The measured blood parameters, heart rates and body temperatures (physiological welfare indictors) of the control ponies in the present study varied over the year but apart from heart rate they did not exceed the critical values given by von Engelhardt and Breves (2000) and Kraft and Dürr (2005). In summer the heart frequency exceeded the limits given for larger horses, but ponies may have higher limits as small animals are known to have higher heart rate frequencies (Stahl, 1967). The horses were able to keep the blood parameters and body temperature within the critical values. This indicates that the ponies adapted adequately and the varying environmental conditions did not affect their welfare.

In contrast the measured blood parameters, heart rate and subcutaneous temperature of the feed restricted ponies showed deviations from the standard values for horses. However, they exhibited no indication of illness during the feeding trial although they appeared more lethargic.

The reduced thyroxin concentrations in the blood indicate that the reduction of the heart rate and the rectal temperature in the restricted fed animals below the critical values was due to a reduced metabolic rate (Cassar-Malek et al., 2001) caused by a reduced heat production due to insufficient feed energy supply. The resting heart rate and the rectal temperature may therefore represent easy measurable indicators that can be used to assess the metabolic adaptations in horse kept outdoors during the winter period.

The blood parameters total protein, β-hydroxybutyrat, total bilirubin and NEFA measured in the feed restricted group showed significant changes compared to the control group. Total bilirubin and total protein exceeded the normal range, indicating beginning diseases especially concerning the liver (von Engelhardt and Breves, 2000; Kraft and Dürr, 2005). The animals' health is highly correlated with its well-being (Flecknell and Molony, 1997) so that ill animals or animals suffering from pain are expected to be restricted in their well-being (Keeling and Jensen, 2002; Knierim, 2002). A reduced feed supply therefore influences the welfare not only by emergence of feelings of hunger but also by health problems.

The experimental ponies showed a highly constant diurnal rhythm in the metabolic rate over the entire year with a nadir before sunrise and a peak after noon. Especially in wintertimes the horses performed a nocturnal hypometabolism. If horses held in extensive housing systems are used by humans, attention should be paid to this rhythm. Furthermore, since the ponies reduced their metabolic rate in winter even under *ad libitum* feed supply an intensive use of the animals (e.g., riding) kept under outdoor conditions is questionable.

In summary, the visual inspection of the ponies and the observation of the behaviour were not sufficient to detect welfare impairment in the restrictedly fed animals under winter outdoor conditions. A combination of behavioural and physiological parameters should be used to assess effects of the housing system on the welfare of the animal. Similarly, Knierim (1998) postulated a combination of several indicators to evaluate the animals' welfare. A visual inspection combined with weekly BCS and measurement of the rectal temperature or heart rate can prevent weight loss and the resulting health problems. Thus, more specific regulations for year round extensive outdoor housing should be given in the guidelines of the respective ministry.

For feed supplemented horses of a robust horse breed such as the Shetland pony the outdoor housing can be an adequate housing system as neither physiological nor behavioural changes indicated welfare impairment. The adaptations of the feed restricted ponies were not sufficient to compensate the lack of feed energy supply so that the health was stressed. Therefore, outdoor housing of horses under winter conditions without feed supplementation can be expected to cause welfare impairment.

Conclusions

The aim of the present study was to examine the adaptation capabilities of a domesticated robust horse breed to extreme environmental conditions and winter feed restriction as found in close to natural housing. On the basis of the results the suitability of a year-round semi-extensive outdoor housing for horses was evaluated.

The diurnal and annual rhythm of locomotor activity and rest in domesticated ponies was similar to that of feral horses and was constant over the year even during feed restriction in winter. An adaptation to the seasonal changes was observed for the quantity of locomotor activity and the total time spent lying.

Decreasing ambient temperatures in winter lead to a reduction in the ponies' metabolic rate that contributed to reduced energy expenditure and lower need for energy intake. This mechanism seems to be part of a genetic predisposition in the ponies as it even appeared with *ad libitum* feed supply.

Results also suggest that ponies dispose of the ability to adapt to changing environmental conditions by the use of a nocturnal hypometabolism that is closely linked to seasonal fluctuations in the photoperiod. This mechanism is similar to that of wild Northern ungulates and allows the animal to minimise the energy demand in late winter and spring when feed availability is low. However, this phenomenon also saves energy in summer. A less pronounced form of the nocturnal hypometabolism is performed daily. This rhythm should be considered in the time of day when the horse is handled. The use of horses by humans e.g., for sports may disturb this diurnal rhythm in metabolic rate, body temperature and activity and may thus have an impact on the well-being and the health of the horse.

The ponies' total water intake (TWI) was very accurately measured with the isotope dilution technique (D_2O). TWI was found to be highly dependent on the ambient temperature and heart rate. The comparison with other herbivores showed that the ponies' TWI fits well into that observed in other domesticated animals. The total body water (TBW) of the Shetland ponies was not affected by season and the developed prediction equation can be used to calculate the TBW of ponies from the body weight.

The study on ten Shetland ponies over a period of one year showed that ponies adapted effectively to extreme environmental conditions by altering behavioural and physiological parameters without health impairment when food supply was sufficient.

The year round outdoor housing of Shetland ponies can therefore be an adequate housing system.

In winter, the restrictedly fed ponies showed more intensive adaptive mechanisms than the fully fed mares. It is suggested, that their thermoregulatory set point was shifted downwards and thereby their zone of thermoneutrality. They exhibited a further reduction in the metabolic rate compared to the *ad libitum* ponies but simultaneously showed first signs of metabolic diseases due to high body fat mobilisation so that a reduction in animal welfare can be assumed. Furthermore, the nocturnal hypometabolism during winter was more pronounced in feed restricted animals. Contrary to speculations, that domestic animals lost their adaptive mechanisms for adaptation to harsh environments compared to wild species, the Shetland ponies showed very effective adaptive mechanisms during food shortage. However, in view of the observed losses in body weight of up to 20% in feed restricted mares, keeping of ponies under extensive outdoor conditions without adequate feed supplementation in winter is expected to cause sincere animal welfare problems even if the ponies adapted similarly to wild horses.

The severe feed restriction of the present study is more likely to be found in close to natural housing of horses (e.g. in landscape management projects). However, an inadequate feed supply can also occur in semi-extensive outdoor housing due to a lack of careful animal inspection. Therefore, more precise guidelines for extensive horse housing should be established since a visual animal inspection is not sufficient. Easily accessible parameters as rectal temperature, heart rate and the body condition scoring can serve as additional indicators to prevent health and welfare problems due to insufficient provision with feed.

The results of the present study indicate that there is a need for more detailed studies on the actual extent of energy savings by the use of the hypometabolism in ponies. As further physiological parameter, the body core temperature variability should be determined, e.g., by the continuous measurement of the rectal or gastrointestinal temperature. Studies in larger horse breeds are of particular interest to evaluate whether the present adaptive capabilities found in Shetland ponies are also present in horse breeds which have been selected for intensive performance and possibly increased metabolic rate.

References

Abdullah, R. and I. R. Falconer (1977). Responses of thyroid activity to feed restriction in goat. *Aust. J. Biol. Sci.* 30, 207-215.

Aganga, A. A., N. N. Umunna, E. O. Oyedipe and P. N. Okoh (1989). Breed differences in water metabolism and body composition of sheep and goats. *J. Agr. Sci.* 113, 255-258.

Al-Ramamneh, D., A. Riek and M. Gerken (2010). Deuterium oxide dilution accurately predicts water intake in sheep and goats. *Animal* 4, 1606-1612.

Alamer, M. (2011). Water requirements and body water distribution in Awassi sheep and Aardi goats during winter and summer seasons. *J. Agr. Sci.* 149, 227-234.

Ali, B. H., T. Hassan and N. Musa (1984). The effect of feed restriction on certain hematological indexes, enzymes and metabolites in Nubian goats. *Comp. Biochem. Physiol A* 79, 325-328.

Arnold, W., T. Ruf and R. Kuntz (2006). Seasonal adjustment of energy budget in a large wild mammal, the Przewalski horse *(Equus ferus przewalskii)* II. Energy expenditure. *J. Exp. Biol.* 209, 4566-4573.

Arnold, W., T. Ruf, S. Reimoser, F. Tataruch, K. Onderscheka and F. Schober (2004). Nocturnal hypometabolism as an overwintering strategy of red deer *(Cervus elaphus). Am. J. Physiol.-Reg. I* 286, R174-R181.

Autio, E. (2008). Loose housing of horses in a cold climate. Doctoral dissertation, University of Kuopio, Kuopio, Finland.

Baetz, A. L. (1976). The effect of fasting on blood constituents in domestic animals. *Ann. Vet. Res.* 7, 105-108.

Bednarski, R. M. and S. M. McGuirk (1986). Bradycardia associated with fasting in cattle. *Vet. Surg.* 15, 458-458.

Beede, D. K. (1993). Water Nutrition and Quality for Dairy Cattle, In: Western Large Herd Management Conference, Las Vegas, Nevada, USA, pp193-205.

Beede, D. K. and R. J. Collier (1986). Potential nutritional strategies for intensively managed cattle during thermal-stress. *J. Anim. Sci.* 62, 543-554.

Berger, A., K. M. Scheibe, K. Eichhorn, A. Scheibe and J. Streich (1999). Diurnal and ultradian rhythms of behaviour in a mare group of Przewalski horse *(Equus ferus przewalskii)*, measured through one year under semi-reserve conditions. *Appl. Anim. Behav. Sci.* 64, 1-17.

Berger, A., K. M. Scheibe, K. Wollenweber, B. Patan, P. Schnitker, C. Herrman and K. D. Budras (2006). Jahresrhythmik von Aktivität, Nahrungsaufnahme, Lebendmasse und Hufentwicklung bei Wild- und Hauspferden in naturnahen Lebensbedingungen. In: Aktuelle Arbeiten zur artgemäßen Tierhaltung 2006. KTBL Schrift 448. KTBL, Darmstadt, pp 137-146.

Bevan, R. M., A. J. Woakes, P. J. Butler and J. P. Croxall (1995). Heart-rate and oxygen-consumption of exercising Gentoo penguins. *Physiol. Zool.* 68, 855-877.

Bianca, W. (1963). Thermoregulatory responses of dehydrated ox to ingestion of cold and warm water in a warm environment. *Res. Vet. Sci.* 5, 75–80.

Blincoe, C. and S. Brody (1955). The influence of ambient temperature, air velocity, radiation intensity, and starvation on thyroid activity and iodine metabolism in cattle. *Mo. Agr. Exp. Sta. Res. Bul.* 576.

Blum, J. W., M. Gingins, P. Vitins and H. Bickel (1980). Thyroid hormone levels related to energy and nitrogen balance during weight loss and regain in adult sheep. *Acta Endocrinol. - Cop.* 93, 440 - 447.

Blum, J. W. and P. Kunz (1981). Effects of fasting on thyroid-hormone levels and kinetics of reverse triiodothyronine in cattle. *Acta Endocrinol.- Cop* 98, 234-239

BMELV - Bundesministerium für Ernährung, Landwirtschaft und Verbraucherschutz (2009). Leitlinien zur Beurteilung von Pferdehaltungen unter Tierschutzgesichtspunkten.
http://www.bmelv.de/SharedDocs/Downloads/Landwirtschaft/Tier/Tierschutz/G utachtenLeitlinien/HaltungPferde.pdf;jsessionid=8D6525E895A29877FD17477 2A7CFFB7F.2_cid181?__blob=publicationFile

BMELV - Bundesministerium für Ernährung, Landwirtschaft und Verbraucherschutz (2006). Tierschutzgesetz. http://www.gesetze-im-internet.de/tierschg/

BMG - Bundesministerium für Gesundheit und Frauen (2010). Mindestanforderungen für die Haltung von Pferden und Pferdeartigen, Schweinen, Rindern, Schafen, Ziegen, Schalenwild, Lamas, Kaninchen, Hausgeflügel, Straußen und Nutzfischen.
http://www.ris.bka.gv.at/GeltendeFassung.wxe?Abfrage=Bundesnormen&Ges etzesnummer=20003820

Bond, J., T. S. Rumsey and B. T. Weinland (1976). Effect of deprivation and reintroduction of feed and water on feed and water-Intake behavior of beef-cattle. *J. Anim. Sci.* 43, 873-878.

Boyd, L. E. (1988). Time budgets of adult Przewalski horses - effects of sex, reproductive status and enclosure. *Appl. Anim. Behav. Sci.* 21, 19-39.

Brockman, R. P. (1979). Roles for insulin and glucagon in the development of ruminant ketosis - Review. *Can. Vet. J.* 20, 121-126.

Broom, D. M. (1991). Animal-welfare - concepts and measurement. *J. Anim. Sci.* 69, 4167-4175.

Broom, D. M. and A. F. Fraser (2007). Domestic Animal Behaviour and Welfare. CAB International, Oxfordshire, UK.

Brosh, A., Y. Aharoni, A. A. Degen, D. Wright and B. Young (1998). Estimation of energy expenditure from heart rate measurements in cattle maintained under different conditions. *J. Anim. Sci.* 76, 3054-3064.

Butler, P. J., J. A. Green, I. L. Boyd and J. R. Speakman (2004). Measuring metabolic rate in the field: the pros and cons of the doubly labelled water and heart rate methods. *Funct. Ecol.* 18, 168-183.

BVET - Bundesamt für Veterinärwesen (2011). Pferde dauernd im Freien halten. http://www.bvet.admin.ch/tsp/02414/03799/index.html?lang=de

Cassar-Malek, I., S. Kahl, C. Jurie and B. Picard (2001). Influence of feeding level during postweaning growth on circulating concentrations of thyroid hormones and extrathyroidal 5 '-deiodination in steers. *J. Anim. Sci.* 79, 2679-2687.

Cassida, K. A., B. A. Barton, R. L. Hough, M. H. Wiedenhoeft, K. Guillard (1994). Feed-intake and apparent digestibility of hay-supplemented brassica diets for lambs. *J. Anim. Sci.* 72, 1623-1629.

Castle, M. E. and T. P. Thomas (1975). Water-Intake of British Friesian cows on rations containing various forages. *Anim. Prod.* 20, 181-189.

Christopherson, R. J., H. W. Gonyou and J. R. Thompson (1979). Effects of temperature and feed-intake on plasma-concentration of thyroid-hormones in beef-cattle. *Can. J. Anim. Sci.* 59, 655-661.

Christopherson, R. J. and P. M. Kennedy (1983). Effect of the thermal environment on digestion in ruminants. *Can. J. Anim. Sci.* 63, 477-496.

Clabough, D. L. and C. R. Swanson (1989). Heart rate spectral analysis of fasting-induced bradycardia of cattle. *Am. J. Physiol.* 257, R1303-1306.

Coenen, M. (2005). Exercise and stress: impact on adaptive processes involving water and electrolytes. *Livest. Prod. Sci.* 92, 131-145.

Dahlborn, K., M. Akerlind and G. Gustafson (1998). Water intake by dairy cows selected for high or low milk-fat percentage when fed two forage to concentrate ratios with hay or silage. *Swed. J. Agr. Res.* 28, 167-176.

Dahlborn, K., B. Samuelsson and K. Svennersten-Sjaunja (1995). Changes in total plasma protein, sodium and aldosterone concentration during milking, feeding, and food deprivation in the Swedish dairy cow. *Swed. J. Agr. Res.* 25, 129-136.

Dahme, E. and E. Weiss (2007). Grundriss der speziellen pathologischen Anatomie der Haustiere. 6th Edition. Enke Verlag Stuttgart, Germany.

DEFRA - Department for Environment, Food and Rual Affairs (2009). Code of Practice for the Welfare of Horses, Ponies, Donkeys and their Hybrids http://www.defra.gov.uk/publications/2011/03/26/code-of-practice-horses-pb13334/

Drivers, J. and S. F. Peek (2008). Diseases of Dairy Cattle. 2nd Edition. Sauders, St. Louis, Missouri, USA.

Duncan, P. (1980). Time-budgets of Camargue horses. 2. Time-budgets of adult horses and weaned sub-adults. *Behaviour* 72, 26-49.

Ekpe, E. D. and R. J. Christopherson (2000). Metabolic and endocrine responses to cold and feed restriction in ruminants. *Can. J. Anim. Sci.* 80, 87-95.

Emmanuel, B. and J. J. Kennelly (1984). Effect of propionic-acid on ketogenesis in lactating sheep fed restricted rations or deprived of food. *J. Dairy Sci.* 67, 344-350.

Engelking, L. R. (1993). Equine Fasting Hyperbilirubinemia. In: Advances in veterinary science and comparative medicine; Animal Models in Liver Research (eds C. E. Cornelius, E. C. Melby and R. R. Marshak), Academic Press Inc, San Diego, USA, pp 115-125.

Fell, L. R. and D. A. Shutt (1989). Behavioral and hormonal responses to acute surgical stress in sheep. *Appl. Anim. Behav. Sci.* 22, 283-294.

Finn, J. P. and B. Tennant (1974). Hepatic encephalopathy in cattle. *Cornell Vet.* 64, 136-153.

Flecknell, P. A. and V. Molony (1997). Pain and Injury. In: Animal Welfare (eds M. C. Appleby and B. O. Hughes), CAB International, Oxon, UK, pp 63-73.

Forbes, J. M. (2003). The multifactorial nature of food intake control. *J. Anim. Sci.* 81, E139–E144.

Frohli, D. and J. W. Blum (1988). Effects of fasting on blood-plasma levels, metabolism and metabolic effects of epinephrine and norepinephrine in steers. *Acta Endocrinol. Cop.* 118, 254-259.

Furll, M. and M. Schafer (1993). Hyperbilirubinemia in ruminants. *Monatsh. Veterinarmed.* 48, 183-187.

Galyean, M. L., R. W. Lee and M. E. Hubbert (1981). Influence of fasting and transit on ruminal and blood metabolites in beef steers. *J. Anim. Sci.* 53, 7-18.

Geiser, F. and T. Ruf (1995). Hibernation versus daily torpor in mammals and birds - physiological variables and classification of torpor patterns. *Physiol. Zool.* 68, 935-966.

Glade, M. J. and T. J. Reimers (1985). Effects of dietary energy supply on serum thyroxine, tri-iodothyronine and insulin concentrations in young horses. *J. Endocrinol.* 104, 93-98.

Gronwall, R. and A. S. Mia (1972). Fasting hyperbilirubinemia in horses. *American J. Dig. Dis.* 17, 473-476.

Guyton, A. C. (1991). Body Temperature, Temperature Regulation and Fever. In: Textbook of medical physiology (eds J. E. Hall and A. C. Guyton), Sauders Company, Philidelphia, USA, Chapter 73.

Heitmann, R. N., S. C. Sensenig, C. K. Reynolds, J. M. Fernandez and D. J. Dawes (1986). Changes in energy metabolite and regulatory hormone concentrations and net fluxes across splanchnic and peripheral-tissues in fed and progressively fasted ewes. *J. Nutr.* 116, 2516-2524.

Hoffman, M. P. and H. L. Self (1972). Factors affecting water consumption by feedlot cattle. *J. Anim. Sci.* 35, 871-876.

Hoffmann, G., S. Rose-Meierhofer and B. Niemann (2009). Water demand of horses and water consumption in horse husbandry. *Tierarztl. Umschau* 64, 438-442.

Holter, J. B. and W. E. Urban (1992). Water partitioning and intake prediction in dry and lactating Holstein cows. *J. Dairy Sci.* 75, 1472-1479.

Iqbal, A., A. M. Cheema and E. R. Kühn (1990). Growth hormone induced stimulation of the T_4 to T_3 conversion in fed and fasting dwarf goats. *Horm. Metab. Res.* 22, 566-568.

Irvine, C. H. G. (1967). Thyroxine secretion rate in horse in various physiological states. *J. Endocrinol.* 39, 313-320.

Kadzere, C. T., M. R. Murphy, N. Silanikove and E. Maltz (2002). Heat stress in lactating dairy cows: a review. *Livest. Prod. Sci.* 77, 59-91.

Kamal, T. H., H. D. Johnson and A. C. Ragsdale (1959). Water consumption in dairy cattle as influenced by environmental temperatures and urine excretion. *J. Dairy Sci.* 42, 926-926.

Katz, M. L. and E. N. Bergman (1969). Hepatic and portal metabolism of glucose free fatty acids and ketone bodies in sheep. *Am. J. Physiol.* 216, 953-960.

Keeling, L. and P. Jensen (2002). Behavioural Disturbances, Stress and Welfare. In: The ethology of domestic animals: an introductory text (eds P. Jensen), CAB International, Wallingford, UK, pp 79-98.

Knierim, U. (1998). Wissenschaftliche Untersuchungsmethoden zur Beurteilung der Tiergerechtheit. In Beurteilung der Tiergerechtheit von Haltungssystemen (eds KTBL-Schrift). 337. 40-50.

Knierim, U. (2002). Basic ethological considerations concerning the assessment of husbandry conditions with regard to farm animal welfare. *Deut. Tierarztl. Woch.* 109, 261-266.

Kolb, E. (1967). Lehrbuch der Physiologie der Haustiere. 2nd Edition. Gustav Fischer Verlag, Jena, Germany.

Kraft, W. and U. M. Dürr (2005). Klinische Labordiagnostik in der Tiermedizin. 6th Edition. Schattauer, Stuttgart, New York.

Kristula, M. A. and S. M. McDonnell (1994). Drinking-water temperature affects consumption of water during cold weather in ponies. *Appl. Anim. Behav. Sci.* 41, 155-160.

Lainez, M. M. and L. C. Hsia (2004). Effects of season, housing and physiological stage on drinking and other related behavior of dairy cows (Bos taurus). *Asian-Aust. J. Anim. Sci.* 17, 1417-1429.

Lamoot, I. and M. Hoffmann (2004). Do season and habitat influence the behaviour of Haffinger mares in a coastal dune area? *Belg. J. Zool.* 134, 97-103.

Looper, L. M. and D. N. Waldner. (2007). Water for dairy cattle. New Mexico State University. http://aces.nmsu.edu/pubs/_d/D-107.pdf

Lundgren, R. G. and H. D. Johnson (1964). Effects of temperature and feed intake on thyroxine disappearance rates in cattle. *J. Anim. Sci* 23, 28-31.

Macfarlane, W. V. and B. Howard (1972). Comparative water and energy economy of wild and domestic mammals. *Sym. Zool. S.* 31, 261-296.

Macfarlane, W. V., B. Howard and R. J. H. Morris (1966). Water metabolism of merino sheep shorn during summer. *Aust. J. Agr. Res.* 17, 219-225.

Maloiy, G. M. O. (1973). Water metabolism of East African ruminants in arid and semi-arid regions. *J. Anim. Breed. Genet.* 90, 219–228.

Matsuyama, S., S. Ohkura, T. Ichimaru, K. Sakurai, H. Tsukamura, K. Maeda and H. Okamura (2004). Simultaneous observation of the GnRH pulse generator activity and plasma concentrations of metabolites and insulin during fasting and subsequent refeeding periods in Shiba goats. *J. Reprod. Develop.* 50, 697-704.

McDowell, R. E. (1972). Improvement of Livestock Production in Warm Climates. W. H. Freeman and Co., San Francisco, USA.

McDowell, R. E. and J. R. Weldy (1967). Water exchange of cattle under heat stress. In: Proceedings of the 3rd International Biometeorological Congress, London. Pergamin Press, New York pp. 414–424.

McSherry, B. J., J. H. Lumsden, V. E. Valli and J. D. Baird (1984). Hyperbilirubinemia in sick cattle. *Can. J. Comp. Med.* 48, 237-240.

Mench, J. A. and J. G. Manson (1997). Behaviour. In: Animal Welfare (eds M. C. Appleby and B. O. Hughes), CAB International, Oxon, UK, 127-141.

Meyer, H. (1992). Pferdefütterung. 2nd Edition. Verlag Paul Parey, Berlin, Hamburg, Germany.

Meyer, U., W. Stahl and G. Flachowsky (2006). Investigations on the water intake of growing bulls. *Livest. Sci.* 103, 186-191.

Moberg, G. P. and J. A. Mench (2000). The Biology of Animal Stress: Basic Principles and Implications for Animal Welfare. CAB International, Wallingford, UK.

Morgan, K., A. Ehrlemark and K. Sallvik (1997). Dissipation of heat from standing horses exposed to ambient temperatures between -3 degrees C and 37 degrees C. *J. Therm. Biol.* 22, 177-186.

Morris, M. D., H. F. Hintz and Zilversm.Db (1972). Hyperlipoproteinemia in fasting ponies. *J. Lipid Res.* 13, 383-389.

Murphy, M. R. (1992). Symposium - Nutritional factors affecting animal water and waste quality - water metabolism of diary cattle. *J. Dairy Sci.* 75, 326-333.

Naqvi, S. M. K. and A. K. Rai (1991). Effect of fasting on some physiological-responses and blood-constituents in native and crossbred sheep. *Indian J. Anim. Sci.* 61, 985-990.

Nielsen, N. I., K. L. Ingvartsen and T. Leasen (2003). Diurnal variation and the effect of feed restriction on plasma and milk metabolites in TMR-fed dairy. cows. *J. Vet. Med. A.* 50, 88-97.

NRC (1964). Nutrient Requirements of Sheep. National Academy Press, Washington D.C., USA.

NRC (1981). Effect of Environment on Nutrient Requirements of Domestic Animals. National Academic Press, Washington D.C., USA.

NRC (1985). Nutrient Requirements of Sheep. National Academy Press, Washington D.C., USA.

NRC (2000). Nutrient Requirements of Beef Cattle. National Academy Press, Washington, USA.

Ott, E. A. (2005). Influence of temperature stress on the energy and protein metabolism and requirements of the working horse. *Livest. Prod. Sci.* 92, 123-130.

Pell, J. M. and E. N. Bergman (1983). Cerebral metabolism of amino-acids and glucose in fed and fasted sheep. *Am. J. Physiol.* 244, E282-E289.

Porzig, E. and H. H. Sambraus (1991). Nahrungsaufnahmeverhalten landwirtschaftlicher Nutztiere. Deutscher Landwirtschaftsverlag Berlin GmbH, Berlin, Germany.

Pothoven, M. A. and D. C. Beitz (1975). Changes in fatty-acid synthesis and lipogenic enzymes in adipose-tissue from fasted and fasted-refed steers. *J. Nutr.* 105, 1055-1061.

Puchala, R., I. Tovar-Luna, T. Sahlu, H. C. Freetly and L. Goetsch (2009). Technical Note: The relationship between heart rate and energy expenditure in growing crossbred Boer and Spanish wethers. *J. Anim. Sci.* 87, 1714-1721.

Renecker, L. A. and R. J. Hudson (1985). Telemetered heart-rate as an index of energy-expenditure in moose *(Alces-alces). Comp. Biochem. Physiol. A.* 82, 161-165.

Rule, D. C., D. C. Beitz, G. Deboer, R. R. Lyle, A. H. Trenkle and J. W. Young (1985). Changes in hormone and metabolite concentrations in plasma of steers during a prolonged fast. *J. Anim. Sci.* 61, 868-875.

Rumsey, T. S. and J. Bond (1976). Cardiorespiratory patterns, rectal temperature, serum electrolytes and packed cell-volume in beef-cattle deprived of feed and water. *J. Anim. Sci.* 42, 1227-1238.

Scheunert, A. and A. Trautmann (1987). Lehrbuch der Veterinär-Physiologie. 7th Edition. Verlag Paul Parey, Berlin Hamburg, Germany.

Seifi, H. A., H. Gray, M. Mohri and A. Badiee (2002). Hyperlipidemia in Caspian miniature horses: Effect of undernutrition. *J. Equine Vet. Sci.* 22, 205-207.

Signer, C., T. Ruf and W. Arnold (2011). Hypometabolism and basking: the strategies of Alpine ibex to endure harsh over-wintering conditions. *Funct. Ecol.* 25, 537-547.

Silanikove, N. (2000). The physiological basis of adaptation in goats to harsh environments. *Small Ruminant Res.* 35, 181-193.

Silanikove, N. and A. Tadmor (1989). Rumen volume, saliva flow-rate, and systemic fluid homeostasis in dehydrated cattle. *Am. J. Physiol.* 256, R809-R815.

Silver, H., N. F. Colovos, J. B. Holter and H. H. Hayes (1969). Fasting metabolism of white-tailed deer. *J. Wildlife Manage.* 33, 490-498.

Singer, C., T. Ruf and W. Arnold (2011). Hypometabolism and basking: the strategies of Alpine ibex to endure over-wintering conditions. *Funct. Ecol.* 25, 537 - 547.

Sjaastad, O. V., K. Hove and O. Sand (2003). Physiology of Domestic Animals. Scandinavian Veterinary Press, Oslo, Norway.

Stahl, W. R. (1967). Scaling of respiratory variables in mammals. *J. Appl. Physiol.* 22, 453-460.

Staufenbiel, R. (2004). Stoffwechselerkrankungen. In: Tiergesundheits- und Tierkrankheitslehre (eds W. Busch, W. Methling and W. M. Anmselgruber), Parey Verlag, Stuttgart, pp 334-363.

Sticker, L. S., D. L. Thompson, L. D. Bunting, J. M. Fernandez and C. L. Depew (1995). Dietary-protein and or energy restriction in mares - plasma-glucose, insulin, nonesterified fatty-acid, and urea nitrogen responses to feeding, glucose, and epinephrine. *J. Anim. Sci.* 73, 136-144.

Terlouw, E. M., G. P. Schouten and J. Ladewig (1997). Physiology. In: Animal Welfare (eds M. C. Appleby and B. O. Hughes), CAB International, Oxon, UK, pp 314-158.

Trenkle, A. (1976). Estimates of kinetic-parameters of growth-hormone metabolism in fed and fasted calves and sheep. *J. Anim. Sci.* 43, 1035-1043.

TVT - Tierärztliche Vereinigung für Tierschutz e.V. (2004). Positionspapier zu den "Leitlinien zur Beurteilung von Pferdehaltungen unter Tierschutzgesichtspunkten". http://www.tierschutz-tvt.de/21.html?&0=

TVT - Tierärztliche Vereinigung für Tierschutz e.V. (2006). Rinder und Pferde in Landschaftspflege- und Naturentwicklungsprojekten. http://www.tierschutz-tvt.de/merkblaetter.html#c6

Vallentine, J. F. (2001). Grazing Management. Academic Press, San Diego, USA.

Veterinär- u. Lebensmittelüberwachungsamt Ennepe-Ruhr-Kreis. (2011) Merkblatt: Ganzjährige Weidehaltung von Pferden. http://www.enkreis.de/fileadmin/civserv/5954008/forms/Merkblatt_Ganzjaehrige_Weidehaltung_Pferde.pdf

von Engelhardt, W. and G. Breves (2000). Physiologie der Haustiere. 3rd Edition. Enke Verlag, Stuttgart, Germany.

Ward, J. R., D. M. Henricks, T. C. Jenkins and W. C. Bridges (1992). Serum hormone and metabolite concentrations in fasted young bulls and steers. *Dom. Anim. Endocrin.* 9, 97-103.

Westra, R. and R. J. Christopherson (1976). Effects of cold on digestibility, retention time of digesta, reticulum motility and thyroid-hormones in sheep. *Can. J. Ani. Sci.* 56, 699-708.

Willen, S. (2004). Tierbezogene Indikatoren zur Beurteilung der Tiergerechtheit in der Milchviehhaltung - methodische Untersuchungen und Beziehungen zum Haltungssystem. Doctoral dissertation, Tierärztliche Hochschule Hannover, Hannover, Germany.

Winchester, C. F. and M. J. Morris (1956). Water intake rates of cattle. *J. Anim. Sci.* 15, 722-740.

Yang, Y. T. and R. L. Baldwin (1973). Lipolysis in isolated cow adipose cells. *J. Dairy Sci.* 56, 366.

Zeitler-Feicht, M. (2001). Handbuch Pferdeverhalten. Ursache, Therapie und Prophylaxe. 2nd Edition. Ulmer-Verlag, Stuttgart, Germany.

Acknowledgements

I would like to thank:

Prof. Dr. Martina Gerken for accepting the topic, for enabling the formation of an experimental herd of ponies and for her continuous support.

Prof. Dr. Dr. Matthias Gauly for taking over the co-referee.

Prof. Dr. Josef Troxler for acting as third examiner.

Dr. Alexander Riek for proof reading my publications and his help with the statistical analysis.

Jürgen Dörl for taking care of the ponies and for being very helpful in technical problems.

Dieter Daniel and the trainees for taking care of the ponies when Jürgen was not there.

The "Mensa Truppe" for various hours of funny discussions during the lunch breaks- not only about academic topics.

Birgit Sohnrey for help and advice during the analysis in the lab.

Michael Beihsner for his critical and helpful comments and the veterinary supervision of the animals.

The team of the "Wissenschaftliche Werkstatt" for giving me a helping hand with technical difficulties.

Vivian Gabor for always being a nice colleague and for the interesting discussions about horse topics.

My husband Cort for his continous encouragement and motivation during all the time.

My parents and my brother for their continous support troughout the years of my education.

Last but not least my experimental ponies ☺

Lebenslauf

Name: Brinkmann, geb. Mann

Vorname: Lea

Geburtsdatum: 22. November 1982

Geburtsort: Göttingen, Deutschland

Schulen

1988 – 1992:   Grundschule Waake, 37136 Waake

1992 – 1994:   Orientierungsstufe Weende Nord, 37075 Göttingen

1994 – 2002:   Otto-Hahn-Gymnasium, 37081 Göttingen

Studium

2003 – 2006   Bachelor Studium Agrarwissenschaften, Fachrichtung Tierproduktion, Georg-August Universität Göttingen, 37073 Göttingen

2006 – 2008   Master Studium Agrarwissenschaften, Fachrichtung Tierproduktion, Georg-August Universität Göttingen, 37073 Göttingen

2008 – 2012   Promotion zum Doktor der Agrarwissenschaften, Georg-August Universität Göttingen, 37073 Göttingen

Seit 4/2012   Postdoc, Department für Nutztierwissenschaften, Georg-August Universität Göttingen, 37073 Göttingen

Praktika

2002   Fachtierarzt für Pferde, Dr. Gremmes, 37127 Jühnde

2004   Landwirtschaftlicher Betrieb mit Pferdezucht Familie Heemke, 27308 Kreepen/Kirchlinteln

2005   Landwirtschaftliches Betriebsgründungs- und beratungsbüro (LBB), 37075 Göttingen

2007   Tierarztpraxis Michael Beihsner, 37136 Ebergötzen

Göttingen, Juni 2012